William Elwood Byerly

Elements of the Differential Calculus

With Examples and Applications

William Elwood Byerly

Elements of the Differential Calculus
With Examples and Applications

ISBN/EAN: 9783337277307

Printed in Europe, USA, Canada, Australia, Japan

Cover: Foto ©berggeist007 / pixelio.de

More available books at **www.hansebooks.com**

ELEMENTS

OF THE

DIFFERENTIAL CALCULUS,

WITH

EXAMPLES AND APPLICATIONS.

A TEXT BOOK

BY

W. E. BYERLY, Ph.D.,
ASSISTANT PROFESSOR OF MATHEMATICS IN HARVARD UNIVERSITY.

BOSTON, U.S.A.:
PUBLISHED BY GINN & COMPANY.
1898.

PREFACE.

The following book, which embodies the results of my own experience in teaching the Calculus at Cornell and Harvard Universities, is intended for a text-book, and not for an exhaustive treatise.

Its peculiarities are the rigorous use of the Doctrine of Limits as a foundation of the subject, and as preliminary to the adoption of the more direct and practically convenient infinitesimal notation and nomenclature; the early introduction of a few simple formulas and methods for integrating; a rather elaborate treatment of the use of infinitesimals in pure geometry; and the attempt to excite and keep up the interest of the student by bringing in throughout the whole book, and not merely at the end, numerous applications to practical problems in geometry and mechanics.

I am greatly indebted to Prof. J. M. Peirce, from whose lectures I have derived many suggestions, and to the works of Benjamin Peirce, Todhunter, Duhamel, and Bertrand, upon which I have drawn freely.

W. E. BYERLY.

CAMBRIDGE, *October, 1879.*

TABLE OF CONTENTS.

CHAPTER I.
INTRODUCTION.

Article.		Page
1.	Definition of *variable* and *constant*	1
2.	Definition of *function* and *independent variable*	1
3.	Symbols by which functional dependence is expressed	2
4.	Definition of *increment*. Notation for an increment. An increment may be positive or negative	2
5.	Definition of the *limit* of a variable	3
6.	Examples of *limits* in Algebra	3
7.	Examples of *limits* in Geometry	4
8.	The fundamental proposition in the *Theory* of *Limits*	5
9.	Application to the proof of the theorem that the area of a circle is one-half the product of the circumference by the radius	5
10.	Importance of the clear conception of a *limit*	6
11.	The velocity of a moving body. *Mean velocity; actual velocity at any instant; uniform velocity; variable velocity*	6
12.	Actual velocity easily indicated by aid of the *increment* notation	7
13.	Velocity of a falling body	7
14.	The direction of the tangent at any point of a given curve. Definition of *tangent* as limiting case of *secant*	8
15.	The *inclination of a curve to the axis of* X easily indicated by the aid of the *increment* notation	8
16.	The inclination of a parabola to the axis of X	9
17.	Fundamental object of the Differential Calculus	10

CHAPTER II.
DIFFERENTIATION OF ALGEBRAIC FUNCTIONS.

18.	Definition of *derivative*. Derivative of a *constant*	11
19.	General method of finding the derivative of any given function. General formula for a derivative. Examples	11

Article.	Page.
20. Classification of *functions*	12
21. Differentiation of the *product of a constant and the variable;* of a *power of the variable*, where the exponent is a positive integer	13
22. Derivative of a *sum of functions*	14
23. Derivative of a *product of functions*	15
24. Derivative of a *quotient of functions.* Examples	17
25. Derivative of a *function of a function* of the variable	18
26. Derivative of a *power* of the variable where the exponent is *negative* or *fractional.* Complete set of formulas for the differentiation of Algebraic functions. Examples	19

CHAPTER III.

APPLICATIONS.

Tangents and Normals.

27. Direction of *tangent* and *normal* to a plane curve	22
28. Equations of *tangent* and *normal. Subtangent. Subnormal. Length* of tangent. *Length* of normal. Examples	23
29. Derivative may sometimes be found by solving an equation. Examples	25

Indeterminate Forms.

30. Definition of *infinite* and *infinitely great*	26
31. Value of a function corresponding to an infinite value of the variable	26
32. Infinite value of a function corresponding to a particular value of the variable	27
33. The expressions $\frac{0}{0}$, $\frac{\infty}{\infty}$, and $0 \times \infty$, called indeterminate forms. When definite values can be attached to them	28
34. Treatment of the form $\frac{0}{0}$. Examples	28
35. Reduction of the forms $\frac{\infty}{\infty}$ and $0 \times \infty$ to the form $\frac{0}{0}$	30

Maxima and Minima of a Continuous Function.

36. *Continuous* change. *Continuous function*	31
37. If a function increases with the increase of the variable, its derivative is positive; if it decreases, negative	31
38. Value of derivative shows rate of increase of function	32
39. Definition of *maximum* and *minimum* values of a function	32

Article.		Page.
40.	*Derivative zero* at a maximum or a minimum	33
41.	Geometrical illustration	33
42.	Sign of derivative near a zero value shown by the value of its own derivative	34
43.	Derivatives of different orders	34
44.	Numerical example	34
45.	Investigation of a minimum	35
46.	Case where the third derivative must be used. Examples	35
47.	General rule for discovering maxima and minima. Examples	36
48.	Use of auxiliary variables. Examples	38
49.	Examples	39

Integration.

50.	Statement of the problem of finding the *distance* traversed by a falling body, given the *velocity*	41
51.	Statement of the problem of finding the *area* bounded by a given curve	41
52.	Statement of the problem of finding the length of an arc of a given curve	42
53.	*Integration. Integral*	44
54.	*Arbitrary constant* in integration	44
55.	Some formulas for direct integration	44
56.	Solution of problem stated in Article 50	45
57.	Example under problem stated in Article 51. Examples	46
58.	Examples under problem stated in Article 52	48

CHAPTER IV.

TRANSCENDENTAL FUNCTIONS.

59.	Differentiation of $\log x$ requires the investigation of the limit of $\left(1+\dfrac{1}{m}\right)^m$	49
60.	Expansion of $\left(1+\dfrac{1}{m}\right)^m$ by the Binomial Theorem	50
61.	Proof that the limit in question is the sum of a well-known series	50
62.	This series is taken as the base of the natural system of logarithms. Computation of its numerical value	52
63.	Extension of the proof given above to the cases where m is not a positive integer	53
64.	Differentiation of $\log x$ completed	54
65.	Differentiation of a^x. Examples	55

DIFFERENTIAL CALCULUS.

Trigonometric Functions.

66. *Circular* measure of an angle. Reduction from *degree* to *circular* measure. Value of the *unit* in circular measure . . . 57
67. Differentiation of $\sin x$ requires the investigation of the limit $\dfrac{\sin \Delta x}{\Delta x}$ and $\dfrac{1-\cos \Delta x}{\Delta x}$ 57
68. Investigation of these limits 58
69. Differentiation of the Trigonometric Functions. Examples . 59
70. *Anti-* or *inverse* Trigonometric Functions 60
71. Differentiation of the Anti-Trigonometric Functions. Examples 60
72. *Anti-* or *inverse* notation. Differentiation of *anti-* functions in general 61
73. The derivative of y with respect to x, and the derivative of x with respect to y, are reciprocals. Examples 62

CHAPTER V.

INTEGRATION.

74. Formulas for *direct* integration 65
75. Integration by *substitution.* Examples 66
76. If fx can be integrated, $f(a+bx)$ can always be integrated. Examples 67
77. $\int_x \dfrac{1}{\sqrt{(a^2-x^2)}}$. Examples 67
78. $\int_x \dfrac{1}{\sqrt{(a^2+x^2)}}$. Example 68
79. *Integration by parts.* Examples 69
80. $\int_x \sin^2 x$. Examples 69
81. Use of *integration* by *substitution* and *integration by parts* in combination. Examples 70
82. Simplification by an *algebraic transformation.* Examples . . . 71

Applications.

83. Area of a segment of a circle; of an ellipse; of an hyperbola . 72
84. Length of an arc of a circle 74
85. Length of an arc of a parabola. Example 75

CHAPTER VI.

CURVATURE.

86. *Total* curvature; *mean* curvature; *actual* curvature. Formula for actual curvature 77

Article.
87. To find *actual curvature* conveniently, an indirect method of differentiation must be used 77
88. The derivative of z with respect to y is the quotient of the derivative of z with respect to x by the derivative of y with respect to x 78
89. Reduced formula for *curvature*. Examples 78
90. *Osculating circle*. *Radius of curvature*. *Centre of curvature* . 81
91. Definition of *evolute*. Formulas for evolute 82
92. Evolute of a parabola 83
93. Reduced formulas for *evolute*. Example 85
94. Evolute of an ellipse. Example 85
95. Every normal to a curve is tangent to the evolute 87
96. Length of an arc of *evolute*. 88
97. Derivation of the name evolute. Involute 88

CHAPTER VII.

THE CYCLOID.

98. Definition of the cycloid 90
99. Equations of cycloid referred to the base and a tangent at the lowest point as axes. Examples 90
100. Equations of the cycloid referred to vertex as origin. Examples 92
101. Statement of properties of cycloid to be investigated 93
102. Direction of tangent and normal. Examples 93
103. Equations of tangent and normal. Example 94
104. Subtangent. Subnormal. Tangent. Normal 94
105. Curvature. Examples 95
106. Evolute of cycloid 96
107. Length of an arc of cyloid 97
108. Area of cycloid. Examples 98
109. Definition and equations of *epicycloid* and *hypocycloid*. Examples 99

CHAPTER VIII.

PROBLEMS IN MECHANICS.

110. Formula for *velocity* in terms of distance and time 102
111. *Acceleration*. Example. Differential equations of motion . . 102
112. Two principles of mechanics taken for granted 103
113. Problem of a body falling freely near the earth's surface . . . 103

Article.	Page.
114. Second method of integrating the equations of motion in the case of a falling body	104
115. Motion down an *inclined plane*	106
116. Motion of a body sliding down a chord of a vertical circle. Example	107
117. Problem of a body falling from a distance toward the earth. *Velocity* of fall. *Limit* of *possible velocity.* *Time* of fall. Examples	108
118. Motion down a smooth curve. Examples	112
119. Time of descent of a particle from any point of the arc of an inverted cycloid to the vertex. *Cycloidal pendulum.* *Tautochrone*	113
120. A problem for practice	116

CHAPTER IX.

DEVELOPMENT IN SERIES.

121. Definition of *series.* *Convergent* series. *Divergent* series. *Sum* of series	117
122. Example of a series	117
123. Function obtained by integration from one of its derivatives. Series suggested	118
124. Development of $f(x_0+h)$ into a series. Determination of the coefficients on the assumption that the development is possible. Examples	119
125. An error committed in taking a given number of terms as equivalent to the function developed	122
126. Lemma	122
127. Error determined	123
128. Second form for *remainder.* *Taylor's Theorem*	125
129. Examples of use of expression for *remainder.* *Test* for the possibility of developing a function	126
130. Development of log $(1+x)$	127
131. The *Binomial Theorem.* Investigation of the cases in which the ordinary development holds for a negative of fractional value of the exponent. Example	129
132. *Maclaurin's Theorem*	132
133. Development of a^x, e^x, and e	133
134. Development of sin x and cos x	134
135. Development of $\sin^{-1} x$ and $\tan^{-1} x$. Examples	134
136. The investigation of the *remainder* in Taylor's Theorem often omitted. Examples	136

Article.	Page.
137. *Leibnitz's Theorem* for *Derivatives* of *a Product* 136	
138. Development of tan x. Example 137	

Indeterminate Forms.

139. Treatment of *indeterminate forms* by the aid of Taylor's Theorem. The form $\frac{0}{0}$. Example 138
140. Special consideration of the case where the form $\frac{0}{0}$ occurs for an infinite value of the variable 139
141. The form $\frac{\infty}{\infty}$; special consideration of the cases where its *true value* is zero or infinite 141
142. Reduction of the forms ∞^0, 1^∞, 0^0 to forms already discussed. Examples 142

Maxima and Minima.

143. Treatment of *maxima and minima* by Taylor's Theorem . . . 145
145. Generalization of the investigation in the preceding article. Examples 146

CHAPTER X.

INFINITESIMALS.

146. Definition of *infinitesimal* 149
147. *Principal infinitesimal*. *Order* of an infinitesimal. Examples . 149
148. Determination of the order of an infinitesimal. Examples . . 150
149. Infinitesimal increments of a function and of the variable on which it depends are of the same order 151
150. Lemma. Expression for the the coördinates of points of a curve by the aid of an auxiliary variable 152
151. Lemma 153
152. Lemma 153
153. Geometrical example of an infinitesimal of the *second order* . 154
154. In determining a tangent the secant line can be replaced by a line infinitely near 155
155. Tangent at any point of the *pedal* of a given curve 156
156. The locus of the foot of a perpendicular let fall from the focus of an ellipse upon a tangent. Example 157
157. Tangent at any point of the locus of a point cutting off a given distance on the normal to a given curve 158
158. Tangent to the locus of the vertex of an angle of constant magnitude, circumscribed about a given curve. Example . 159

DIFFERENTIAL CALCULUS.

Article. Page.
159. The substitution of one *infinitesimal* for another 160
160. Theorem concerning the *limit of the ratio* of two infinitesimals 160
161. Theorem concerning the *limit of the sum* of infinitesimals . . 161
162. If two infinitesimals differ from each other by an infinitesimal of higher order, the limit of their ratio is unity 162
163. Direction of a tangent to a parabola 163
164. Area of a sector of a parabola 164
165. The limit of the ratio of an infinitesimal arc to its chord is unity 165
166. Rough use of infinitesimals 166
167. Tangent to an ellipse. Examples 167
168. The area of a segment of a parabola. Examples 168
169. New way of regarding the cycloid 169
170. *Tangent* to the cycloid 170
171. *Area* of the cycloid 171
172. *Length* of an arc of the cycloid 172
173. *Radius of curvature* of the cycloid 174
174. *Evolute* of the cycloid 176
175. Examples 177
176. The *brachistochrone, or curve of quickest descent* a cycloid . . 177

CHAPTER XI.

DIFFERENTIALS.

177. In obtaining a *derivative* the *increment* of the function may be replaced by a simpler infinitesimal. Application to the *derivative of an area;* to the *derivative of an arc* 183
178. Definition of *differential* 185
179. Differential notation for a *derivative.* A derivative is the actual ratio of two differentials 185
180. Advantage of the differential notation 185
181. Formulas for differentials of functions. Examples 186
182. The differential notation especially convenient in dealing with problems in *integration.* Numerical example 187
183. Integral regarded as the *limit of a sum of differentials. Definite integral* 188
184. An area regarded as the limit of a sum of infinitesimal rectangles. Example 189
185. Definition of *centre of gravity.* The centre of gravity of a parabola 190

Differentials of Different Orders.

Article.		Page.
186.	Definition of the *order* of a differential	192
187.	Relations between differentials and derivatives of different orders. Assumption that the differential of the independent variable is constant	193
188.	Expression for the second derivative in terms of differentials when no assumption is made concerning the differential of the independent variable.	194
189.	Differential expression for the *radius of curvature*	194
190.	*Finite differences* or *increments* of different orders	194
191.	Any infinitesimal increment differs from the differential of the same order by an infinitesimal of higher order	196
192.	Lemma	197
193.	Proof of statement in Article 191	197

CHAPTER XII.

FUNCTIONS OF MORE THAN ONE VARIABLE.

Partial Derivatives.

194.	Illustration of a *function of two variables*	199
195.	Definition of *partial derivative* of a function of several variables. Illustration.	199
196.	*Successive partial derivatives*	200
197.	In obtaining successive partial derivatives the order in which the differentiations occur is of no consequence	200
198.	*Complete differential* of a function of two variables. Example .	202
199.	Use of partial derivatives in obtaining *ordinary* or *complete* derivatives. Example.	203
200.	Special case of Article 199. Examples	204
201.	Use of partial derivatives in finding successive complete derivatives. Example	205
202.	Derivative of an implicit function. Examples	206
203.	Second derivative of an implicit function. Examples	207

CHAPTER XIII.

CHANGE OF VARIABLE.

204.	If the *independent variable* is changed, differentials of higher orders than the first must be replaced by more general values.	209
205.	Example.	209

xiv   DIFFERENTIAL CALCULUS.

Article.   Page.
206. Example of the change of both dependent and independent variable at the same time. Example 211
207. Direction of a tangent to a curve in terms of polar coördinates. Examples 213
208. Treatment of the subject of *change of variable* without the use of differentials. Example 214
209. Change of variable when partial derivatives are employed. First Method. Examples 215
210. Second Method. Examples 217
211. Third Method. Examples 218

CHAPTER XIV.

TANGENT LINES AND PLANES IN SPACE.

212. A curve in space is represented by a pair of simultaneous equations 220
213. Equations of *tangent* line to a curve in space. Equation of *normal plane* 220
214. Tangent and normal to *helix* 221
215. Expressions for equation of tangent line to curve in space in terms of partial derivatives. Examples 224
216. *Osculating plane* to a curve in space. Example 225
217. *Tangent plane* to a *surface*. Examples 226

CHAPTER XV.

DEVELOPMENT OF A FUNCTION OF SEVERAL VARIABLES.

218. *Taylor's* and *Maclaurin's* Theorems for *functions of two independent variables*. Example 227
219. Taylor's Theorem for three variables. Example 231
220. *Euler's Theorem* for *homogeneous functions*. Example . . . 232

CHAPTER XVI.

MAXIMA AND MINIMA OF FUNCTIONS OF TWO OR MORE VARIABLES.

221. Essential conditions for the existence of a *maximum* or *minimum* 234
222. Tests for the detection of maxima and minima 235
223. Formulas for maxima and minima 236
224. Examples 236
225. Examples 239

CHAPTER XVII.

THEORY OF PLANE CURVES.

Concavity and Convexity.

Article.		Page.
226.	Tests for *concavity* and *convexity* of plane curves	240
227.	*Points of inflection*	240
228.	Application of *Taylor's* Theorem to the treatment of *convexity* and *concavity* and *points* of inflection	241
229.	Examples	243
230.	*Singular points*	245
231.	Definition of *multiple point; osculating point; cusp; conjugate point; point d'arrêt; point saillant*	245
232.	Test for a *multiple point*	246
233.	Detection of *osculating points, cusps, conjugate points, points d'arrêt,* and *points saillant*	247
234.	Example of a *double point*	247
235.	Example of a *cusp*	249
236.	Example of a *conjugate point*. Examples	250

Contact of Curves.

237.	*Orders* of contact	251
238.	Order of contact indicates closeness of contact	252
239.	*Osculating circle*. Examples	253

Envelops.

240.	An equation may represent a *series* of curves. *Variable parameter. Envelop*	255
241.	Determination of the equation of an *envelop*	255
242.	Example	257
243.	An *evolute* the *envelop* of the *normal*. Examples	257

DIFFERENTIAL CALCULUS.

CHAPTER I.

INTRODUCTION.

1. A *variable quantity*, or simply a *variable*, is a quantity which, under the conditions of the problem into which it enters, is susceptible of an indefinite number of values.

A *constant quantity*, or simply a *constant*, is a quantity which has a fixed value.

For example; in the equation of a circle

$$x^2 + y^2 = a^2,$$

x and y are variables, as they stand for the coördinates of any point of the circle, and so may have any values consistent with that fact; that is, they may have an unlimited number of different values; a is a constant, since it represents the radius of the circle, and has therefore a fixed value. Of course, any given number is a constant.

2. When one quantity depends upon another for its value, so that a change in the second produces a change in the first, the first is called a *function* of the second. If, as is generally the case, the two quantities in question are so related that a change in either produces a change in the other, either may be regarded as a function of the other. The one of which the other is considered a function is called *the independent variable*, or simply *the variable*.

For example; if x and y are two variables connected by the relation
$$y = x^2,$$
we may regard x as the independent variable, and then y will be a function of x, for any change in x produces a corresponding change in its square; or we may regard y as the independent variable, and then x will be a function of y, and from that point of view the relation would be more naturally written
$$x = \sqrt{y}.$$

Again, suppose the relation is
$$y = \sin x,$$
we may either regard y as a function of x, in which case we should naturally write the relation as above, or we may regard x as a function of y, and then we should more naturally express the same relation by
$$x = \sin^{-1} y,$$
i.e., x is equal to the angle whose sine is y.

3. Functional dependence is usually indicated by the letters f, F, and φ. Thus we may indicate that y is a function of x by writing
$$y = fx, \text{ or } y = Fx, \text{ or } y = \varphi x;$$
and in each of these expressions the letter f, F, or φ is not an algebraic quantity, but a mere symbol or abbreviation for the word *function*, and the equation is precisely equivalent to the sentence, y *depends upon* x *for its value, so that a change in the value of* x *will necessarily produce a change in the value of* y.

4. The difference between any two values of a variable is called an *increment* of the variable, since it may be regarded as the amount that must be added to the first value to produce the second. An increment is denoted by writing the letter Δ before the variable in question. Thus the difference between two values of a variable x would be written Δx, Δ being merely a sym-

bol for the word *increment*, and the expression dx representing a single quantity. It is to be noted that as an increment is a difference, it may be either positive or negative.

5. *If a variable* which changes its value according to some law *can be made to approach some fixed, constant value as nearly as we please, but can never become equal to it*, the constant is called the *limit* of the variable under the circumstances in question.

6. For example; the limit of $\frac{1}{n}$, *as* n *increases indefinitely*, is zero; for by making n sufficiently great we can evidently decrease $\frac{1}{n}$ at pleasure, but we can never make it absolutely zero.

The sum of n terms of the geometrical progression 1, $\frac{1}{2}$, $\frac{1}{4}$, $\frac{1}{8}$, &c., is a variable that changes as n changes, and if n is increased at pleasure, the sum will have 2 for its limit; for, by the formula for the sum of a geometrical progression,

$$s = \frac{ar^n - a}{r - 1}.$$

In this case, $\quad s = \dfrac{\frac{1}{2^n} - 1}{\frac{1}{2} - 1} = \dfrac{1 - \frac{1}{2^n}}{\frac{1}{2}}.$

By increasing n, $\frac{1}{2^n}$ can be made as small as we please, but cannot become absolutely zero; the numerator can then be made to approach the value 1 as nearly as we please, and the limit of the value of the fraction is obviously 2.

We say, then, that the limit of the sum of n terms of the progression 1, $\frac{1}{2}$, $\frac{1}{4}$, $\frac{1}{8}$, &c., *as* n *increases indefinitely*, is 2.

In both of these examples the variable increases towards its limit, but remains always less than its limit. This, however, is not always the case. The variable may decrease towards its limit remaining always greater than the limit, or in approaching its limit, it may be sometimes greater and sometimes less than the limit. Take, for example, the sum of n terms of the progression $1, -\frac{1}{2}, \frac{1}{4}, -\frac{1}{8}, \frac{1}{16}$, &c., where the ratio is $-\frac{1}{2}$. Here the limit of the sum *as* n *increases indefinitely*, is $\frac{2}{3}$. Let n start with the value 1 and increase; when

$$n=1, \; s=1,$$

and is greater than the limit $\frac{2}{3}$; when

$$n=2, \; s=\frac{1}{2},$$

and is less than $\frac{2}{3}$, but is nearer $\frac{2}{3}$ than 1 was; when

$$n=3, \; s=\frac{3}{4},$$

and is greater than $\frac{2}{3}$; when

$$n=4, \; s=\frac{5}{8},$$

and is less than $\frac{2}{3}$; and as n increases, the values of s are alternately greater and less than the limit $\frac{2}{3}$, but each value of s is nearer $\frac{2}{3}$ than the value before it.

7. It follows immediately from the definition of a *limit*, that *the difference between a variable and its limit is itself a variable which has zero for its limit*, and in order to prove that a given constant is the limit of a particular variable, it will always suffice to show that the difference between the two has the limit zero.

For example; it is shown in elementary geometry that the difference between the area of any circle and the area of the inscribed or circumscribed regular polygon can be made as small as we please by increasing the number of sides of the polygon, and this difference evidently can never become absolutely zero. The area of a circle is then the limit of the area of the regular inscribed or circumscribed polygon as the number of sides of the polygon is indefinitely increased.

It is also shown in geometry, that the difference between the length of the circumference of a circle and the length of the perimeter of the regular inscribed or circumscribed polygon can be decreased indefinitely by increasing at pleasure the number of sides of the polygon, and this difference evidently can never become zero. The length of the circumference of a circle is then the limit of the length of the perimeter of the regular inscribed or circumscribed polygon as the number of sides of the latter is indefinitely increased.

8. The fundamental proposition in the *theory of limits* is the following

THEOREM. — *If two variables are so related that as they change they keep always equal to each other, and each approaches a limit, their limits are absolutely equal.*

For two variables so related that they are always equal form but a single varying value, as at any instant of their change they are by hypothesis absolutely the same. A single varying value cannot be made to approach at the same time two different constant values as nearly as we please; for, if it could, it could eventually be made to assume a value between the two constants; and, after that, in approaching one it would recede from the other.

9. As an example of the use of this principle, let us prove that the area of a circle is one-half the product of the length of its radius by the length of its circumference.

Circumscribe about the circle any regular polygon, and join its vertices with the centre of the circle, thus dividing it into a set of triangles, each having for its base a side of the polygon, and for its altitude the radius of the circle. The area of each triangle is one-half the product of its base by the radius. The sum of these areas, or the area of the polygon, is one-half the length of the radius by the sum of the lengths of the sides, that is, by the length of the perimeter of the polygon. If A' is the area, and P the perimeter of the polygon, and R the radius of the circle, we have

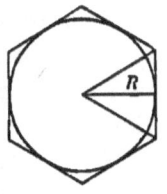

$$A' = \tfrac{1}{2} RP;$$

a relation that holds, no matter what the number of sides of the polygon. A' and $\frac{1}{2}RP$ evidently change as we change the number of sides of the polygon; they are then two variables so related that, as they change, they keep always equal to each other. As the number of sides of the polygon is indefinitely increased, A' has the area of the circle as its limit; P has the circumference of the circle as its limit. Let A be the area and C the circumference of the circle; then the

$$\text{limit } A' = A,$$

and the $\qquad$ limit $\frac{1}{2}RP = \frac{1}{2}RC.$

By the Theorem of Limits these limits must be absolutely equal;

$$\therefore A = \tfrac{1}{2}RC. \qquad\qquad \text{Q.E.D.}$$

10. It is of the utmost importance that the student should have a perfectly clear idea of a *limit*, as it is by the aid of this idea that many of the fundamental conceptions of mechanics and geometry can be most clearly realized in thought.

11. Let us consider briefly the subject of the velocity of a moving body.

The *mean velocity* of a moving body, during any period of time considered, is the quotient obtained by dividing the distance traversed by the body in the given period by the length of the period, the distance being expressed in terms of a unit of length, and the length of the period in terms of some unit of time.

If, for example, a body travels 60 feet in 3 seconds, its mean velocity during that period is said to be $\frac{60}{3}$, or 20; and the body is said to move at the mean rate of 20 feet per second.

The velocity of a moving body is *uniform* when its mean velocity is the same whatever the length of the period considered.

The *actual velocity* of a moving body at any instant, is the *limit* which the body's *mean velocity during the period* immediately succeeding the instant in question approaches as the length of the period is indefinitely decreased. In the case of uniform velocity, the actual velocity at any instant is obviously the same as the actual velocity at any other instant.

If the actual velocity of a moving body is continually changing, the body is said to move with a *variable velocity*.

12. If the law governing the motion of a moving body can be formulated so as to express the distance traversed by the body in any given time as a function of the time, we can indicate the actual velocity at any instant very simply by the aid of the increment notation already explained. Represent the distance by s and the time by t. Then we have

$$s = ft.$$

Suppose we want to find the actual velocity at the end of t_0 seconds. Let Δt be any arbitrary period immediately succeeding the end of t_0 seconds (it can fairly be considered an increment as we really increase the time during which the body is supposed to have moved by Δt seconds), and let Δs be the distance traversed in that period. Then, by definition, the mean velocity during the period Δt is $\dfrac{\Delta s}{\Delta t}$, and the actual velocity desired is the limit approached by this ratio as Δt approaches zero. We shall indicate this by

$$\lim_{\Delta t \doteq 0} \left[\frac{\Delta s}{\Delta t} \right],$$

which is to be read "the limit of Δs divided by Δt, as Δt approaches zero"; the sign $\doteq$ standing for the word *approaches*.

13. Take a numerical example. In the case of a body falling freely in a vacuum near the surface of the earth, the relation connecting the distance fallen with the time is nearly

$$s = 16\,t^2,$$

s being expressed in feet and t in seconds; required, *the actual velocity of a falling body* at the end of t_0 seconds. Let Δt seconds be an arbitrary period immediately after the end of t_0 seconds, then in $t_0 + \Delta t$ seconds

the body would fall $16(t_0 + \Delta t)^2$ feet,

or $16 t_0^2 + 32 t_0 \Delta t + 16 (\Delta t)^2$ feet.

In t_0 seconds it falls $16 t_0^2$ feet, so that in the period Δt in question,

it would fall $16 t_0^2 + 32 t_0 \Delta t + 16 (\Delta t)^2 - 16 t_0^2$ feet,

or $32 t_0 \Delta t + 16 (\Delta t)^2$ feet,

which must therefore be Δs. If v_0 be the required actual velocity,

$$v_0 = \lim_{\Delta t \doteq 0} \left[\frac{\Delta s}{\Delta t} \right], \qquad \text{by Art. 12.}$$

$$\frac{\Delta s}{\Delta t} = \frac{32 t_0 \Delta t + 16 (\Delta t)^2}{\Delta t} = 32 t_0 + 16 \Delta t,$$

and obviously $\lim_{\Delta t \doteq 0} \left[\frac{\Delta s}{\Delta t} \right] = 32 t_0.$

Hence $v_0 = 32 t_0,$

the result required; and in general, the velocity v at the end of t seconds is

$$v = 32 t.$$

14. Let us now consider a geometrical problem: *To find the direction of the tangent* at any point of a given curve.

The *tangent* to a curve, at any given point, is the line with which the secant through the given point and any second point of the curve, tends to coincide as the second point is brought indefinitely near the first. In other words, its position is the *limiting position* of the secant line *as the second point of intersection approaches the first*, i.e., a position that the secant line can be made to approach as nearly as we please, but cannot actually assume.

15. Suppose we have the equation of a curve in rectangular coördinates, and wish to find the angle τ that the tangent at a given point (x_0, y_0) of the curve makes with the axis of X; that is, what is called *the inclination of the curve to the axis of* X.

The equation of the curve enables us to express y in terms of x, that is, as a function of x. We have then

$$y = fx.$$

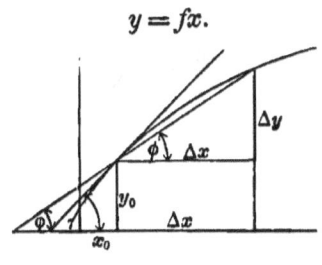

Let
$$x_0 + \Delta x$$

be the abscissa of any second point P of the curve, and

$$y_0 + \Delta y$$

the corresponding ordinate. If φ is the angle which the secant through P_0 and P makes with the axis of X, it is clear from the figure that

$$\tan \varphi = \frac{\Delta y}{\Delta x}.$$

As P approaches P_0, that is, as Δx decreases toward zero, φ evidently approaches τ as its limit, and $\tan \varphi$ of course approaches indefinitely $\tan \tau$. Hence, by the fundamental theorem of limits (Art. 8),

$$\tan \tau = \lim_{\Delta x \doteq 0} \left[\frac{\Delta y}{\Delta x} \right].$$

16. Take a particular example. To find the inclination τ_0 to the axis of X, of the parabola

$$y^2 = 2mx$$

at the point (x_0, y_0) of the curve.

If the abscissa of P is $x_0 + \Delta x$, its ordinate $y_0 + \Delta y$ must be

$$\sqrt{[2m(x_0 + \Delta x)]},$$

as is clear from the equation of the curve, which may be written

$$y = \sqrt{(2mx)}.$$

$$y_0 = \sqrt{(2mx_0)},$$

Δy must be $\quad \sqrt{[2m(x_0+\Delta x)]} - \sqrt{(2mx_0)}.$

$$\frac{\Delta y}{\Delta x} = \frac{\sqrt{[2m(x_0+\Delta x)]} - \sqrt{(2mx_0)}}{\Delta x};$$

or, multiplying numerator and denominator by

$$\sqrt{[2m(x_0+\Delta x)]} + \sqrt{(2mx_0)}$$

to rationalize the numerator,

$$\frac{\Delta y}{\Delta x} = \frac{2m(x_0+\Delta x) - 2mx_0}{\Delta x \{\sqrt{[2m(x_0+\Delta x)]} + \sqrt{(2mx_0)}\}},$$

and $\tan \tau_0 = \underset{\Delta x \doteq 0}{\text{limit}} \left[\dfrac{\Delta y}{\Delta x} \right] = \dfrac{2m}{2\sqrt{(2mx_0)}} = \dfrac{m}{\sqrt{(2mx_0)}} = \dfrac{m}{y_0}.$

At any point (x, y) of the parabola we should have

$$\tan \tau = \frac{m}{y}.$$

At the extremity of the latus rectum, i.e., at the point $\left(\dfrac{m}{2}, m \right)$,

$$\tan \tau = \frac{m}{m} = 1,$$

and $\quad\quad\quad\quad\quad\quad \tau = 45°,$

a familiar property of the parabola.

17. Each of the problems we have just considered has required for its solution the investigation of the limit approached by the ratio of corresponding increments of a function and of the variable on which it depends, *as the increment of the independent variable approaches zero.* Such a limit is called a *derivative,* or a *differential coefficient,* and the study of its form and properties is the fundamental object of the *Differential Calculus.*

CHAPTER II.

DIFFERENTIATION OF ALGEBRAIC FUNCTIONS.

18. If y is a function of x, the *limit of the ratio of an increment of* y *to the corresponding increment of* x, *as the increment of* x *approaches zero*, is called the *derivative of* y *with respect to* x, and is indicated by $D_x y$, D_x being merely an abbreviation for *derivative with respect to* x.* For any particular value of x, this limit, as we shall see, will, in general, have a perfectly definite value; but it will change in value as x changes; that is, the derivative will, in general, be a new function of x.

Since our definition of derivative requires that y should be a function of x, that is, should change when x changes, it follows that a constant can have no derivative; and if we attempt to find the derivative of a constant by the method which we should use if it were a function of x, we shall be led to this same conclusion. Let a be any constant; then the increment produced in a, by giving x any increment, is absolutely 0; the ratio of this increment to the increment of x must then be 0; and as this ratio is always 0, its limit, when we suppose the increment of x to decrease, must be 0. Therefore

$$D_x a = 0. \qquad [1]$$

19. The *general method of finding the derivative of any given function of* x, is immediately suggested by the definition of a derivative. Take two values of x, x_0 and $x_0 + \Delta x$, and find the corresponding values of the given function; the difference between them is obviously the increment of the function, corresponding to the increment Δx of x. The limit of the ratio of the

---

* The names *differential coefficient* and *derived function*, and the notation $\frac{dy}{dx}$ in place of $D_x y$, are also in common use

two increments, as Δx approaches zero, will be the value of the derivative for the particular value x_0 of x, and we may indicate it by $[D_x y]_{x=x_0}$. As x_0 was taken at the start as any value of x, the subscripts may be dropped in the result, and the derivative will then be expressed as an ordinary function of x. The method may be formulated as follows:—

$$[D_x fx]_{x=x_0} = \underset{\Delta x \doteq 0}{\text{limit}} \left[\frac{f(x_0 + \Delta x) - fx_0}{\Delta x} \right]. \qquad [1]$$

The student will observe that, in the problems in Arts. 13 and 16, we have really found $D_t(16 t^2)$ and $D_x(\sqrt{2mx})$ by the method just described.

Examples.

Find

(1) $D_x(20x)$; (2) $D_x(x^3)$; (3) $D_x\left(\dfrac{1}{x}\right)$; (4) $D_x(\sqrt{x})$;

by the general method.

Ans. (1) 20; (2) $3x^2$; (3) $-\dfrac{1}{x^2}$; (4) $\dfrac{1}{2\sqrt{(x)}}$.

20. In order to deal readily with problems into which derivatives enter, it is desirable to work out a complete set of formulas, or rules for finding the derivatives of ordinary functions; and it will be well to begin by roughly classifying functions.

The functions ordinarily considered are:—

(1) *Algebraic Functions:* those in which the only operations performed upon the variable, are the ordinary algebraic operations, namely: Addition, Subtraction, Multiplication, Division, Involution, and Evolution.

Example. $\quad \tfrac{1}{2}x^2 + 3\sqrt[3]{(x-1)}$.

(2) *Logarithmic Functions:* those involving a logarithm of the variable, or of a function of the variable.

Examples. $\quad x \log x$;

$\log(x^2 - ax + b)$.

(3) *Exponential Functions:* those in which the variable, or a function of the variable, appears as an exponent.

Example. $a^{\sqrt{(x^2-x)}}$.

(4) *Trigonometric Functions:*

Example. $\cos x - \sin^2 x$.

21. We shall consider first, the *differentiation** *of Algebraic Functions* of x.

Required $D_x(ax)$ *where a is a constant.*

By the general method (Art. 19),

$$[D_x ax]_{x=x_0} = \underset{\Delta x \doteq 0}{\text{limit}} \left[\frac{a(x_0 + \Delta x) - ax_0}{\Delta x} \right] = \underset{\Delta x \doteq 0}{\text{limit}} \left[\frac{a\Delta x}{\Delta x} \right]$$

$$= \underset{\Delta x \doteq 0}{\text{limit}} [a] = a;$$

$$\therefore D_x(ax) = a. \qquad [1]$$

If $a=1$, this becomes $\quad D_x x = 1.$ $\qquad$ [2]

Required $D_x x^n$ *where n is a positive integer.*

$$[D_x x^n]_{x=x_0} = \underset{\Delta x \doteq 0}{\text{limit}} \left[\frac{(x_0 + \Delta x)^n - x_0^n}{\Delta x} \right].$$

By the Binomial Theorem,

$$(x_0 + \Delta x)^n = x_0^n + nx_0^{n-1}\Delta x + \frac{n(n-1)}{2} x_0^{n-2}(\Delta x)^2 + \ldots + (\Delta x)^n$$

$$\frac{(x_0 + \Delta x)^n - x_0^n}{\Delta x} = nx_0^{n-1} + \frac{n(n-1)}{2} x_0^{n-2}\Delta x + \ldots + (\Delta x)^{n-1}.$$

Each term after the first contains Δx as a factor, and therefore has zero for its limit as Δx approaches zero, so that

$$\underset{\Delta x \doteq 0}{\text{limit}} \left[\frac{(x_0 + \Delta x)^n - x_0^n}{\Delta x} \right] = nx_0^{n-1};$$

$$\therefore [D_x x^n]_{x=x_0} = nx_0^{n-1}.$$

---

* To *differentiate* is to find the derivative.

As x_0 is any value of x, we may drop the subscript, and we have

$$D_x x^n = nx^{n-1}. \qquad [3]$$

22. We shall next consider complex functions composed of two or more functions connected by algebraic operations; the *sum* of several functions, the *product* of functions, the *quotient* of functions.

Required the derivative of $u + v + w$,

where each of the quantities u, v, and w is a function of x.

Let Δx be any increment given to x, and Δu, Δv, and Δw the corresponding increments of u, v, and w. Then, obviously, the increment of the sum $u + v + w$

is equal to $\qquad \Delta u + \Delta v + \Delta w,\qquad$ and we have

$$D_x(u+v+w) = \lim_{\Delta x \doteq 0}\left[\frac{\Delta u + \Delta v + \Delta w}{\Delta x}\right] = \lim_{\Delta x \doteq 0}\left[\frac{\Delta u}{\Delta x} + \frac{\Delta v}{\Delta x} + \frac{\Delta w}{\Delta x}\right]$$

$$= \lim_{\Delta x \doteq 0}\left[\frac{\Delta u}{\Delta x}\right] + \lim_{\Delta x \doteq 0}\left[\frac{\Delta v}{\Delta x}\right] + \lim_{\Delta x \doteq 0}\left[\frac{\Delta w}{\Delta x}\right];$$

but, since Δu and Δx are corresponding increments of the function u and the independent variable x,

$$\lim_{\Delta x \doteq 0}\left[\frac{\Delta u}{\Delta x}\right] = D_x u;$$

in like manner $\qquad \lim_{\Delta x \doteq 0}\left[\frac{\Delta v}{\Delta x}\right] = D_x v,$

and $\qquad \lim_{\Delta x \doteq 0}\left[\frac{\Delta w}{\Delta x}\right] = D_x w;$

hence $\qquad D_x(u+v+w) = D_x u + D_x v + D_x w. \qquad [1]$

It is easily seen that the same proof in effect may be given, whatever the number of terms in the sum, and whether the connecting signs are plus or minus. So, using sum in the sense of algebraic sum, we can say. *the derivative with respect to* x *of the sum of a set of functions of* x *is equal to the sum of the derivatives of the separate functions.*

23. *Required, the derivative of the product* uv, *where u and v are functions of x.*

Let x_0, u_0, and v_0 be corresponding values of x, u, and v; let Δx be an increment given to x, and Δu and Δv the corresponding increments of u and v. Then,

$$[D_x(uv)]_{x=x_0} = \underset{\Delta x \doteq 0}{\text{limit}} \left[\frac{(u_0 + \Delta u)(v_0 + \Delta v) - u_0 v_0}{\Delta x} \right].$$

$$(u_0 + \Delta u)(v_0 + \Delta v) - u_0 v_0 = u_0 \Delta v + v_0 \Delta u + \Delta u \Delta v$$

and $\quad [D_x(uv)]_{x=x_0} = \underset{\Delta x \doteq 0}{\text{limit}} \left[\dfrac{u_0 \Delta v + v_0 \Delta u + \Delta u \Delta v}{\Delta x} \right]$

$$= \underset{\Delta x \doteq 0}{\text{limit}} \left[u_0 \frac{\Delta v}{\Delta x} \right] + \underset{\Delta x \doteq 0}{\text{limit}} \left[v_0 \frac{\Delta u}{\Delta x} \right] + \underset{\Delta x \doteq 0}{\text{limit}} \left[\frac{\Delta u \Delta v}{\Delta x} \right].$$

u_0 does not change as Δx changes, and

$$\underset{\Delta x \doteq 0}{\text{limit}} \left[\frac{\Delta v}{\Delta x} \right] = [D_x v]_{x=x_0};$$

so that $\quad \underset{\Delta x \doteq 0}{\text{limit}} \left[u_0 \dfrac{\Delta v}{\Delta x} \right] = u_0 [D_x v]_{x=x_0};$

and in like manner

$$\underset{\Delta x \doteq 0}{\text{limit}} \left[v_0 \frac{\Delta u}{\Delta x} \right] = v_0 [D_x u]_{x=x_0}.$$

$\dfrac{\Delta u \Delta v}{\Delta x}$ may be written $\Delta u \dfrac{\Delta v}{\Delta x}$ or $\Delta v \dfrac{\Delta u}{\Delta x}$. Let us consider

$$\underset{\Delta x \doteq 0}{\text{limit}} \left[\Delta u \frac{\Delta v}{\Delta x} \right].$$

As Δx approaches 0, Δu, being the corresponding increment of the function u, will also approach 0; and the product $\Delta u \dfrac{\Delta v}{\Delta x}$ will approach 0 as its limit, if $\dfrac{\Delta v}{\Delta x}$ approaches any definite value;

that is, if $D_x v$ has a definite value. It is, however, perfectly conceivable that $\dfrac{\Delta v}{\Delta x}$ may increase indefinitely as Δx approaches zero, instead of having a definite limit; and, in that case, if $\dfrac{\Delta v}{\Delta x}$ should increase rapidly enough to make up for the simultaneous decrease in Δu, the product $\Delta u \dfrac{\Delta v}{\Delta x}$ would not approach zero.

We shall see, however, as we investigate all ordinary functions, that their derivatives have in general fixed definite values for any given value of the independent variable; but, until this is established, we can only say, that

$$\underset{\Delta x \doteq 0}{\text{limit}}\left[\Delta u\,\frac{\Delta v}{\Delta x}\right]=0$$

when $\dfrac{\Delta v}{\Delta x}$ or $\dfrac{\Delta u}{\Delta x}$ has a definite limit, as $\Delta x \doteq 0$; that is, when

$$[D_x v]_{x=x_0} \text{ or } [D_x u]_{x=x_0}$$

has a definite value. With this proviso, we can say,

$$[D_x(uv)]_{x=x_0} = v_0 [D_x u]_{x=x_0} + u_0 [D_x v]_{x=x_0};$$

or, dropping subscripts,

$$D_x(uv) = u D_x v + v D_x u. \qquad [1]$$

Divide through by uv, and we have the equivalent form,

$$\frac{D_x(uv)}{uv} = \frac{D_x u}{u} + \frac{D_x v}{v}. \qquad [2]$$

If we have a product of three factors, as uvw, we can represent the product of two of them, say vw, by z, and we have

$$\frac{D_x(uvw)}{uvw} = \frac{D_x(uz)}{uz} = \frac{D_x u}{u} + \frac{D_x z}{z}.$$

But

$$\frac{D_x z}{z} = \frac{D_x(vw)}{vw} = \frac{D_x v}{v} + \frac{D_x w}{w};$$

CHAP. II.] DIFFERENTIATION. 17

$$\therefore \frac{D_x(uvw)}{uvw} = \frac{D_x u}{u} + \frac{D_x v}{v} + \frac{D_x w}{w}. \qquad [3]$$

This process may be extended to any number of factors, and we shall have the *derivative of a product of functions divided by the product equal to the sum of the terms obtained by dividing the derivative of each function by the function itself.*

24. Required the derivative of the quotient $\frac{u}{v}$, where u and v are functions of x. Employing our usual notation, we have

$$\left[D_x\left(\frac{u}{v}\right)\right]_{x=x_0} = \lim_{\Delta x \doteq 0} \left[\frac{\frac{u_0 + \Delta u}{v_0 + \Delta v} - \frac{u_0}{v_0}}{\Delta x}\right];$$

but

$$\frac{u_0 + \Delta u}{v_0 + \Delta v} - \frac{u_0}{v_0} = \frac{v_0 \Delta u - u_0 \Delta v}{v_0^2 + v_0 \Delta v},$$

and dividing by Δx, we have

$$\left[D_x\left(\frac{u}{v}\right)\right]_{x=x_0} = \lim_{\Delta x \doteq 0} \left[\frac{v_0 \frac{\Delta u}{\Delta x} - u_0 \frac{\Delta v}{\Delta x}}{v_0^2 + v_0 \Delta v}\right]$$

$$= \frac{v_0 [D_x u]_{x=x_0} - u_0 [D_x v]_{x=x_0}}{v_0^2},$$

and dropping subscripts,

$$D_x\left(\frac{u}{v}\right) = \frac{v D_x u - u D_x v}{v^2}. \qquad [1]$$

EXAMPLES.

Find

(1) $D_x[x^3 + x - \sqrt{(x)}]$; (2) $D_x[x^2 \sqrt{(x)}]$; (3) $D_x \frac{\sqrt{(x)}}{x^4}$.

Ans.

(1) $3x^2 + 1 - \frac{1}{2\sqrt{(x)}}$; (2) $\frac{x^2}{2\sqrt{(x)}} + 2x\sqrt{(x)}$; (3) $-\frac{7}{2x^4\sqrt{(x)}}$.

(4) Find, by Art. 24, [1], $D_x\left(\dfrac{1}{x}\right)$.

(5) Deduce $D_x x^n$ from last part of Art. 23.

25. If the quantity to be differentiated is a function of a function of x, it is always theoretically possible, by performing the indicated operations, to express it directly as a function of x, and then to find its derivative by the ordinary rules; but it can usually be more easily treated by the aid of a formula which we shall proceed to establish.

Required, $D_x fy$, y *being itself a function of* x. Let x_0 and y_0 be corresponding values of x and y; let Δx be any increment given to x, and Δy the corresponding increment of y; then

$$[D_x fy]_{x=x_0} = \lim_{\Delta x \doteq 0}\left[\frac{f(y_0 + \Delta y) - fy_0}{\Delta x}\right],$$

and this can be written

$$\lim_{\Delta x \doteq 0}\left[\frac{f(y_0 + \Delta y) - fy_0}{\Delta y} \cdot \frac{\Delta y}{\Delta x}\right].$$

As Δx and Δy are corresponding increments, they approach zero together; hence

$$\lim_{\Delta x \doteq 0}\left[\frac{f(y_0 + \Delta y) - fy_0}{\Delta y}\right]$$

is the same as

$$\lim_{\Delta y \doteq 0}\left[\frac{f(y_0 + \Delta y) - fy_0}{\Delta y}\right],$$

which is equal to $[D_y fy]_{y=y_0}$.

$$\therefore [D_x fy]_{x=x_0} = [D_y fy]_{y=y_0} \cdot [D_x y]_{x=x_0};$$

or, dropping subscripts,

$$D_x fy = D_y fy \cdot D_x y. \qquad [1]$$

This gives immediately, as extensions of Art. 21, [1] and [3],

$$D_x(ay) = a D_x y, \qquad\qquad D_x y^n = n y^{n-1} D_x y.$$

26. Art. 21, [3] can now be readily extended to the case where n *is any number positive or negative, whole or fractional.*

Let n be a negative whole number $-m$, m of course being a positive whole number. Let

$$y = x^n = x^{-m},$$

then we want $D_x y$. Multiplying both members of

$$y = x^{-m}$$

by x^m, we have $\quad x^m y = 1.$

Since $x^m y$ is a constant, its derivative with respect to x must be zero; but by Art. 23, [1] and Art. 21, [3],

$$D_x[x^m y] = x^m D_x y + y D_x x^m = x^m D_x y + m x^{m-1} y$$

m being a positive integer;

$$\therefore x^m D_x y + m x^{m-1} y = 0,$$

and $\quad D_x y = -m x^{-1} y = -m x^{-m-1} = n x^{n-1}.$ Q.E.D.

Let n be any fraction $\dfrac{p}{q}$ where p and q are integers either positive or negative. As before, let

$$y = x^n = x^{\frac{p}{q}}; \qquad \text{required } D_x y.$$

Clearing $\quad y = x^{\frac{p}{q}} \quad$ of radicals,

we have $\quad y^q = x^p;$

and since the two members are equal functions of x, their derivatives must be equal;

$$D_x y^q = D_x x^p,$$

or $\quad q y^{q-1} D_x y = p x^{p-1},$

and $\quad D_x y = \dfrac{p}{q} \cdot \dfrac{x^{p-1}}{y^{q-1}} = \dfrac{p}{q} \cdot \dfrac{x^{p-1}}{(x^{\frac{p}{q}})^{q-1}} = \dfrac{p}{q} \cdot x^{\frac{p}{q}-1} = n x^{n-1}.$ Q.E.D.

The formula, $$D_x x^n = nx^{n-1},$$

Art. 21, [3], holds, then, whatever the value of n.

EXAMPLE.

Prove Art. 24, [1] by the aid of Art. 21, [3] and Art. 23, [1], regarding $\dfrac{u}{v}$ as a product, namely uv^{-1}.

By the aid of these formulas,

$$D_x a = 0;$$ [1]

$$D_x ax = a;$$ [2]

$$D_x x = 1;$$ [3]

$$D_x x^n = nx^{n-1};$$ [4]

$$D_x (u+v+w) = D_x u + D_x v + D_x w;$$ [5]

$$D_x (uv) = u D_x v + v D_x u;$$ [6]

$$D_x \left(\frac{u}{v}\right) = \frac{v D_x u - u D_x v}{v^2};$$ [7]

$$D_x (fy) = D_y (fy) \cdot D_x y;$$ [8]

any algebraic function, no matter how complicated, may be differentiated.

EXAMPLES.

Find $D_x u$ in each of the following cases: —

(1) $u = m + nx.$ *Ans.* $D_x u = n.$

(2) $u = (a + bx) x^3.$ *Ans.* $D_x u = (4bx + 3a) x^2.$

(3) $u = \sqrt{(x^2 + a^2)}$.

Solution: $u = \sqrt{(x^2 + a^2)} = (x^2 + a^2)^{\frac{1}{2}}$.

Let $\quad y = x^2 + a^2$,

then $\quad u = y^{\frac{1}{2}}$.

$$D_x u = D_x y^{\frac{1}{2}} = D_y y^{\frac{1}{2}} \cdot D_x y \quad \text{by Art. 26, [8],}$$

$$D_y y^{\frac{1}{2}} = \tfrac{1}{2} y^{-\frac{1}{2}} \quad \text{by Art. 26, [4],}$$

$$D_x y = 2x;$$

$$\therefore D_x u = xy^{-\frac{1}{2}} = x(x^2 + a^2)^{-\frac{1}{2}} = \frac{x}{\sqrt{(x^2 + a^2)}}.$$

(4) $u = \left(\dfrac{x}{1+x}\right)^n$. $\qquad$ Ans. $D_x u = \dfrac{nx^{n-1}}{(1+x)^{n+1}}$.

(5) $u = \dfrac{x^2}{(a + x^3)^2}$. $\qquad$ Ans. $D_x u = \dfrac{2x(a - 2x^3)}{(a + x^3)^3}$.

(6) $u = (1 + x)\sqrt{(1 - x)}$. $\qquad$ Ans. $D_x u = \dfrac{1 - 3x}{2\sqrt{(1 - x)}}$.

(7) $u = \dfrac{x}{x + \sqrt{(x^2 + 1)}}$. $\qquad$ Ans. $D_x u = \dfrac{2x^2 + 1}{\sqrt{(x^2 + 1)}} - 2x$.

(8) $u = \sqrt{\left(\dfrac{1 - \sqrt{(x)}}{1 + \sqrt{(x)}}\right)}$. $\qquad$ Ans. $D_x u = -\dfrac{1}{2(1 + \sqrt{x})\sqrt{(x - x^2)}}$.

(9) $u = \sqrt{\left(\dfrac{1 + x}{1 - x}\right)}$. $\qquad$ Ans. $D_x u = \dfrac{1}{(1 - x)\sqrt{(1 - x^2)}}$.

(10) $u = \dfrac{\sqrt{(1 + x^2)} + \sqrt{(1 - x^2)}}{\sqrt{(1 + x^2)} - \sqrt{(1 - x^2)}}$.

$\qquad\qquad$ Ans. $D_x u = -\dfrac{2}{x^3}\left[1 + \dfrac{1}{\sqrt{(1 - x^4)}}\right]$.

CHAPTER III.

APPLICATIONS.

Tangents and Normals.

27. We have shown, in Art. 15, that the angle τ, made with the axis of x by the tangent at any given point of a plane curve, when the equation

$$y = fx$$

of the curve, referred to rectangular axes, is known, may be found by the relation

$$\tan \tau = \lim_{\Delta x \doteq 0} \left[\frac{\Delta y}{\Delta x} \right],$$

where Δy and Δx are corresponding increments of y and x, the coördinates of a point of the curve. If the point be (x_0, y_0), we have, then,

$$\tan \tau_0 = [D_x y]_{x=x_0}.$$

At any point (x,y) $\qquad \tan \tau = D_x y.$ $\hfill [1]$

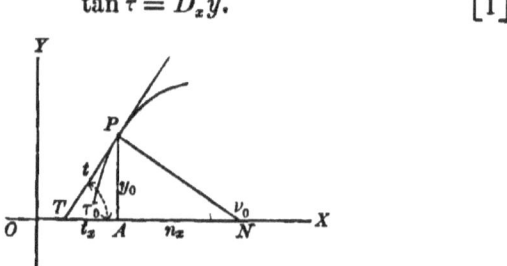

A line perpendicular to any tangent, and passing through the point of contact of the tangent with the curve, is called the *normal* to the curve at that point. If ν_0 be the angle which the normal at the point (x_0, y_0) makes with the axis of X, then it is evident from the figure, that

$$\nu_0 = 90° + \tau_0,$$

and from trigonometry,

$$\tan \nu_0 = -\cot \tau_0 = -\frac{1}{[D_x y]_{x=x_0}}.$$

Of course, for any point (x, y)

$$\tan \nu = -\frac{1}{D_x y}. \qquad [2]$$

28. Since the tangent at (x_0, y_0) passes through (x_0, y_0), and makes an angle τ_0 with the axis of x, its equation will be, by analytic geometry,

$$y - y_0 = \tan \tau_0 (x - x_0);$$

or, since $\tan \tau_0 = [D_x y]_{x=x_0}$,

$$y - y_0 = [D_x y]_{x=x_0} (x - x_0). \qquad [1]$$

In like manner, the equation of the normal at (x_0, y_0) is found to be

$$y - y_0 = -\frac{1}{[D_x y]_{x=x_0}} (x - x_0). \qquad [2]$$

The distance *from* the point of intersection of the tangent with the axis of X *to* the foot of the ordinate of the point of contact, is called the *subtangent*, and is denoted by t_x.

The distance *from* the foot of the ordinate of the point *to* the intersection of the normal with the axis of X, is called the *subnormal*, and is denoted by n_x.

In the figure, TA and AN are respectively the *subtangent* and *subnormal*, corresponding to the point (x_0, y_0) of the curve.

Obviously
$$\frac{y_0}{t_x} = \tan \tau_0 = [D_x y]_{x=x_0},$$

and $\quad \dfrac{y_0}{n_x} = \tan(180° - \nu_0) = -\tan \nu_0 = \dfrac{1}{[D_x y]_{x=x_0}}$;

hence for the point (x_0, y_0),

$$t_x = \dfrac{y_0}{[D_x y]_{x=x_0}} \quad , \quad n_x = y_0 [D_x y]_{x=x_0}.$$

The distance from the intersection of the tangent with the axis of X to the point of contact is sometimes called *the length of the tangent*, and may be denoted by t.

The distance from the point at which the normal is drawn to the point where the normal crosses the axis of X is sometimes called *the length of the normal*, and may be denoted by n.

It is easily seen from the figure, that

$$t = \sqrt{(y_0^2 + t_x^2)},$$

and $\quad n = \sqrt{(y_0^2 + n_x^2)}$;

hence $\quad t = y_0 [D_x y]_{x=x_0}^{-1} \sqrt{(1 + [D_x y]_{x=x_0}^2)}$,

and $\quad n = y_0 \sqrt{(1 + [D_x y]_{x=x_0}^2)}$.

For any point (x, y), our formulas become

$$t_x = \dfrac{y}{D_x y} ; \tag{3}$$

$$n_x = y D_x y ; \tag{4}$$

$$t = y [D_x y]^{-1} \sqrt{(1 + [D_x y]^2)} ; \tag{5}$$

$$n = y \sqrt{(1 + [D_x y]^2)}. \tag{6}$$

EXAMPLES.

(1) Show that the inclination of a straight line to the axis of X is the same at every point of the line; i.e., prove $\tan \tau$ constant.

(2) Show that the subnormal in a parabola

$$y^2 = 2mx$$

is constant, and that the subtangent is always twice the abscissa of the point of contact of the tangent.

(3) Find what point of the parabola must be taken in order that the inclination of the tangent to the axis of X may be 45°.

29. If the equation of the curve cannot be readily thrown into the form $\quad y = fx,$

$D_x y$ *may be found by differentiating both members with respect to* x *and solving the resulting equation algebraically, regarding* $D_x y$ *as the unknown quantity.*

For example; required the equation of the tangent to a circle at the point (x_0, y_0) of the curve. The equation of a circle is

$$x^2 + y^2 = r^2,$$

r being constant. Differentiating with respect to x, we have, by Art. 26, [8],

$$2x + 2y D_x y = 0.$$

Solving, $\qquad D_x y = -\dfrac{2x}{2y} = -\dfrac{x}{y}.$

$$[D_x y]_{x=x_0} = -\dfrac{x_0}{y_0},$$

and by Art. 28, [1], the required equation is

$$y - y_0 = -\dfrac{x_0}{y_0}(x - x_0);$$

or, clearing of fractions,

$$y_0 y - y_0^2 = -x_0 x + x_0^2,$$

$$x_0 x + y_0 y = x_0^2 + y_0^2;$$

but (x_0, y_0) is on the curve, hence

$$x_0^2 + y_0^2 = r^2,$$

and we have

$$x_0 x + y_0 y = r^2,$$

the familiar form of the equation.

EXAMPLES.

(1) Find the equation of the normal at (x_0, y_0) in the circle; of the tangent and the normal at (x_0, y_0) in the ellipse and the hyperbola referred to their axes and centre.

(2) Find at what angle the curve $y^2 = 2ax$

cuts the curve $\quad x^3 - 3axy + y^3 = 0.$

$$\textit{Ans. } \cot^{-1} \sqrt[3]{4}.$$

(3) Show that in the curve $x^{\frac{2}{3}} + y^{\frac{2}{3}} = a^{\frac{2}{3}}$

the length of that part of the tangent intercepted between the axes is constant and equal to a.

Indeterminate Forms.

30. When, under the conditions of the problem, *the value of a variable quantity is supposed to increase indefinitely*, that is, *to increase without limit*, so that the variable can be made greater than any *assigned* value, the variable is called an *infinitely great* quantity or simply an *infinite* quantity, and is usually represented by the symbol ∞. Since infinite quantities are variables, they will usually present themselves to us either as values of the independent variable or as values of a function.

31. *By a value of a function corresponding to an infinite value of the variable*, we shall mean the *limit approached by the value of the function as the value of the variable increases indefinitely.*

Thus, if $\quad y = fx,$

and y approaches the value a as its limit as x increases indefinitely, the value of y corresponding to the value ∞ of x is a, or as we shall say, for the sake of brevity,

$$y = a \text{ when } x = \infty.$$

Since $\frac{1}{x}$ approaches 0 as its limit as x increases indefinitely, we say

$$\frac{1}{x} = 0 \text{ when } x = \infty,$$

or, more briefly,

$$\frac{1}{\infty} = 0.$$

If, as the variable increases indefinitely the function instead of approaching a limit, itself increases indefinitely, we shall say

$$y = \infty \text{ when } x = \infty,$$

meaning, of course, y increases indefinitely when x increases indefinitely.

32. *If, as the variable approaches indefinitely a particular value, the function increases without limit*, we say that *the function is infinite for that particular value of the variable*. For example; as the angle φ approaches the value 90°, its tangent increases indefinitely, and by taking φ sufficiently near 90°, $\tan \varphi$ can be made greater than any assigned value. So we say

$$\tan \varphi = \infty \text{ when } \varphi = 90°,$$

or, more briefly still, $\tan 90° = \infty.$

Again, $\frac{1}{x}$ increases indefinitely as x approaches zero; so we say

$$\frac{1}{x} = \infty \text{ when } x = 0,$$

or simply

$$\frac{1}{0} = \infty.$$

The student can easily convince himself, by a little consideration, that our definition of *infinite* is entirely consistent with the ordinary use of the term in algebra, trigonometry, and analytic geometry.

33. The expressions, $\frac{0}{0}$, $\frac{\infty}{\infty}$, and $0 \times \infty$, are called indeterminate forms; and as they stand, each of them may have any value whatever; for consider them in turn: — By the ordinary definition of a quotient as "a quantity that, multiplied by the divisor, will produce the dividend," $\frac{0}{0}$ may be anything, as any quantity multiplied by 0 will produce 0.

So, too, $\frac{\infty}{\infty}$ may have any value, as obviously any given quantity multiplied by a quantity that increases without limit will give a quantity increasing without limit.

That $0 \times \infty$ is indeterminate is not quite so obvious; for, since zero multiplied by any quantity gives 0, it would seem that zero multiplied by a quantity which increases indefinitely must still give zero, as is indeed the case; and it is only when $0 \times \infty$ presents itself as the limiting value of a product of two variable factors, one of which decreases as the other increases, that we can regard it as indeterminate. In this case the value of the product will depend upon the relative decrease and increase of the two factors, and not merely upon the fact that one approaches zero as the other increases indefinitely.

It is only when $\frac{0}{0}$, $\frac{\infty}{\infty}$, and $0 \times \infty$ occur *in particular problems* as *limiting forms*, that we are able to attach definite values to them.

34. Each of the forms $\frac{\infty}{\infty}$ and $0 \times \infty$, as we shall soon see, can be easily reduced to the form $\frac{0}{0}$, and this form we shall now proceed to study.

If $fx = 0$ and $Fx = 0$ when $x = a$,

the fraction $\frac{fx}{Fx}$, which is, of course, a new function of x, assumes the indeterminate form $\frac{0}{0}$ when $x = a$, and the limit approached by the fraction as x approaches a is called the *true value* of the fraction when $x = a$, and can generally be readily determined.

By hypothesis $\quad fa = 0$ and $Fa = 0$,

hence we can throw $\frac{fx}{Fx}$ into the form $\frac{fx - fa}{Fx - Fa}$, for in so doing we are subtracting 0 from the numerator and 0 from the denominator of the fraction. Again, we can divide numerator and denominator by $x - a$ without changing the value of the fraction;

therefore
$$\frac{fx}{Fx} = \frac{\dfrac{fx - fa}{x - a}}{\dfrac{Fx - Fa}{x - a}},$$

and the *true value* of

$$\left[\frac{fx}{Fx}\right]_{x=a} = \lim_{x \doteq a}\left[\frac{fx}{Fx}\right] = \lim_{x \doteq a} \left[\frac{\dfrac{fx - fa}{x - a}}{\dfrac{Fx - Fa}{x - a}}\right].$$

But $x - a$, being the difference between two values of the variable, is an increment of x; $fx - fa$, being the difference between the values of fx which correspond to x and a, is the corresponding increment of the function, hence

$$\lim_{x \doteq a}\left[\frac{fx - fa}{x - a}\right] = [D_x fx]_{x=a};$$

and in the same way it can be shown that

$$\lim_{x \doteq a}\left[\frac{Fx - Fa}{x - a}\right] = [D_x Fx]_{x=a};$$

wherefore the true value of $\frac{fx}{Fx}$ when $x = a$ is $\frac{[D_x fx]_{x=a}}{[D_x Fx]_{x=a}}$. We have then only to *differentiate numerator and denominator, and substitute in the new fraction* a *for* x, *in order to get the true value* required. It may happen that the new fraction is also indeterminate when $x = a$; if so, we must apply to it the same process that we applied to the original fraction.

The student will observe that this method is based upon the supposition that $\quad fa = 0 \text{ and } Fa = 0,$

so that it is only in this case that we have established the relation

$$\left[\frac{fx}{Fx}\right]_{x=a} = \frac{[D_x fx]_{x=a}}{[D_x Fx]_{x=a}}.$$

EXAMPLES.

Find the true values of the following expressions:—

(1) $\left[\dfrac{x-1}{x^n-1}\right]_{x=1}.$ *Ans.* $\dfrac{1}{n}.$

(2) $\left[\dfrac{x^4 + 3x^3 - 7x^2 - 27x - 18}{x^4 - 3x^3 - 7x^2 + 27x - 18}\right]_{x=3}.$ *Ans.* 10.

(3) $\left[\dfrac{x^{\frac{3}{2}} - 1 + (x-1)^{\frac{3}{2}}}{(x^2-1)^{\frac{3}{2}} - x + 1}\right]_{x=1}.$ *Ans.* $-\dfrac{3}{2}.$

(4) $\left[\dfrac{x^{\frac{3}{2}} - 1 + (x-1)^{\frac{3}{2}}}{\sqrt{(x^2-1)}}\right]_{x=1}.$ *Ans.* 0.

(5) $\left[\dfrac{1 - \sqrt{(1-x)}}{\sqrt{(1+x)} - \sqrt{(1+x^2)}}\right]_{x=0}.$ *Ans.* 1.

(6) $\left[\dfrac{\sqrt{(x)} - \sqrt{(a)} + \sqrt{(x-a)}}{\sqrt{(x^2-a^2)}}\right]_{x=a}.$ *Ans.* $\dfrac{1}{\sqrt{(2a)}}.$

(7) $\left[\dfrac{x^3 - 3x + 2}{x^4 - 6x^2 + 8x - 3}\right]_{x=1}.$ *Ans.* $\infty.$

35. If fx and Fx both increase indefinitely as x approaches the value a, or, as we say for the sake of brevity, if

$fa = \infty$ and $Fa = \infty$,

we can determine the true value of $\left[\dfrac{fx}{Fx}\right]_{x=a}$ by first throwing the fraction $\dfrac{fx}{Fx}$ into the equivalent form $\dfrac{\frac{1}{Fx}}{\frac{1}{fx}}$, which assumes the form $\dfrac{0}{0}$ when $x = a$, and may be treated by the method just described.

If $\qquad fx = 0$ and $Fx = \infty$ when $x = a$,

the true value of $[fx.Fx]_{x=a}$ can be determined by throwing $fx.Fx$ into the equivalent form $\dfrac{fx}{\frac{1}{Fx}}$, which assumes the form $\dfrac{0}{0}$ when $x = a$.

Maxima and Minima of a Continuous Function.

36. A *variable* is said to *change continuously* from one value to another when it changes *gradually* from the first value to the second, *passing through all the intermediate values*.

A *function* is said to be *continuous* between two given values of the variable, when it has a *single finite value* for every value of the variable between the given values, and *changes gradually* as the variable passes from the first value to the second.

37. If the function is *increasing* as the variable *increases*, the increment Δy, produced by adding to x a positive increment Δx, will be positive; $\dfrac{\Delta y}{\Delta x}$ will therefore be positive, and $\lim\limits_{\Delta x \doteq 0}\left[\dfrac{\Delta y}{\Delta x}\right]$ will also be positive; that is, $D_x y$ *will be positive*.

If a function *decreases* as the variable *increases*, the increment Δy, produced by giving x a positive increment Δx, will be negative; $\dfrac{\Delta y}{\Delta x}$ will therefore be negative, and $\lim\limits_{\Delta x \doteq 0}\left[\dfrac{\Delta y}{\Delta x}\right]$ will also be negative; that is, $D_x y$ *will be negative*.

Since $D_x y$, being, as we have seen, itself a function of x, may happen to be positive for some values of x and negative for others, it would seem that the same function may be sometimes increasing and sometimes decreasing as the variable increases, and this is often obviously the case. For example; $\sin \varphi$ increases as φ increases, while φ is passing through the values between $0°$ and $90°$; but it decreases as φ increases, while φ is passing through the values between $90°$ and $180°$.

38. Not only does any particular value of the derivative of a function show by its *sign* whether the function is *increasing* or *decreasing* with the increase of the variable, but it shows by its *numerical magnitude* the *rate* at which the function is changing in comparison with the change in the variable as the latter is passing through the corresponding value.

For example; when $x = 2$, $D_x x^2$ or $2x$ equals 4, and this shows that when x increasing is passing through the value 2, its square is increasing four times as fast.

For if Δx and Δy are corresponding increments of the variable and the function, starting from a particular value x_0 of x, $\dfrac{\Delta y}{\Delta x}$ may be regarded as the mean rate of change in y compared with the change in x, and $\lim\limits_{\Delta x \doteq 0} \left[\dfrac{\Delta y}{\Delta x} \right]$ will then show the actual rate of change at the instant x passes through the value x_0.

39. If, as the variable increases, the function *increases up to a certain value and then decreases*, that value is called a *maximum* value of the function.

If, as the variable increases, the function *decreases to a certain value and then increases*, that value is called a *minimum* value of the function.

In these definitions of maximum and minimum values, the variable is supposed to *increase continuously*.

As a maximum value is merely a value greater than the values immediately before and immediately after it, *a function may have several different maximum values;* and, for a like reason, *it may have several different minimum values.* If

$$y = fx$$

be the equation of the curve in the figure, the ordinates y_1 and y_2 are *maximum* values of y, y_3 and y_4 are *minimum* values of y.

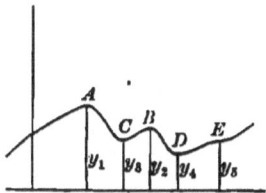

40. *In the following discussion we shall suppose throughout that the variable continually increases.* Then, as at a maximum value, the function by definition changes from increasing to decreasing, its derivative must, by Art. 37, be changing from a positive to a negative value; and if the derivative is a continuous function of the variable in the neighborhood of the value in question, it can change from a positive to a negative value only by passing through the value zero.

Since, at a minimum value, the function by definition changes from decreasing to increasing, its derivative must be changing from a negative to a positive value, and must therefore be passing through the value zero, provided that it is a continuous function of the variable in the neighborhood of the value in question.

41. Confining ourselves for the present to the case *where the derivative is a continuous function*, we can say then, that if y is a function of x, any value x_0 *of* x *corresponding to a maximum or a minimum value of* y *must make* $D_x y$ *zero.* This can also be seen from the figure of Art. 39. For, at the points A, B, C, and D, the tangent to the curve is parallel to the axis of X, and therefore at each of these points $D_x y$, which is, by Art. 27, the

tangent of the inclination of the curve to the axis, must equal zero.

Of course it does not follow from the argument just presented, that every value of x that makes $D_x y = 0$ must correspond either to a maximum or a minimum value of y; and it is evident, from the figure just referred to, that, at the point E, the tangent is parallel to the axis of X, and $D_x y$ is zero, although y_5 is neither a maximum nor a minimum.

42. In order to ascertain the precise nature of the value of y corresponding to a given value of x which makes $D_x y$ zero, we need to know the *sign of* $D_x y$ *for values of* x *just before and just after the value in question*, and this can generally be determined *by noting the value of the derivative of* $D_x y$, which we can always find, as $D_x y$ itself is a function of x, and can be differentiated.

43. $D_x(D_x y)$ is called the *second derivative of* y *with respect to* x, and is denoted by $D_x^2 y$. $D_x(D_x^2 y)$ is called the third derivative of y with respect to x, and is denoted by $D_x^3 y$; and in general, if n is any positive whole number, $D_x(D_x^{n-1} y)$ is called the nth derivative of y with respect to x, and is denoted by $D_x^n y$.

44. *Example.* Required the nature of the value of $x^3 - x^2$ corresponding to the value 0 of x.

Let
$$y = x^3 - x^2 :$$
$$D_x y = 3x^2 - 2x,$$
$$D_x^2 y = 6x - 2;$$
$$[D_x y]_{x=0} = 0,$$
$$[D_x^2 y]_{x=0} = -2.$$

Since $D_x^2 y$ is negative when $x = 0$, $D_x y$ must have been decreasing as x passed through the value zero, and as

$$[D_x y]_{x=0} = 0$$

$D_x y$ must have been positive before $x = 0$, and negative after $x = 0$; therefore, y must have been increasing before $x = 0$, and decreasing after $x = 0$, and must consequently have a maximum value when $x = 0$. To confirm our conclusion, let us find the values of $x^3 - x^2$ when $x = -.1$, when $x = 0$, and when $x = .1$:

$$[x^3 - x^2]_{x=-.1} = -.011,$$
$$[x^3 - x^2]_{x=0} = 0,$$
$$[x^3 - x^2]_{x=.1} = -.009;$$

and the value corresponding to $x = 0$ is the greatest of the three.

45. If $\quad [D_x y]_{x=x_0} = 0$

and $\quad [D_x^2 y]_{x=x_0} > 0,$

$D_x y$ must have been increasing as x passed through the value x_0; and, therefore, since $D_x y = 0$ when $x = x_0$, it must have been negative before $x = x_0$ and positive after $x = x_0$: y then must have been decreasing before $x = x_0$ and increasing after $x = x_0$, and so must be a minimum when $x = x_0$.

46. If $\quad [D_x y]_{x=x_0} = 0$

and $\quad [D_x^2 y]_{x=x_0} = 0,$

we must find the value of $D_x^3 y$ before we can decide on the nature of y_0. Suppose $\quad [D_x y]_{x=x_0} = 0,$

$$[D_x^2 y]_{x=x_0} = 0,$$

and $\quad [D_x^3 y]_{x=x_0} < 0.$

As $[D_x^3 y]_{x=x_0}$ is negative, $D_x^2 y$ must have been decreasing as x passed through the value x_0, and being 0 when $x = x_0$, must have been positive before and negative after. $D_x y$ therefore must have been increasing before $x = x_0$ and decreasing after;

and as $$[D_x y]_{x=x_0} = 0,$$

it must have been negative both before and after $x = x_0$. The function y, then, must have been decreasing both before and after $x = x_0$, and y_0 is neither a maximum nor a minimum.

EXAMPLES.

(1) Show that if $[D_x y]_{x=x_0} = 0$,
$[D_x^2 y]_{x=x_0} = 0$,
and $[D_x^3 y]_{x=x_0} > 0$,
y_0 is neither a maximum nor a minimum.

(2) If $[D_x y]_{x=x_0} = 0$,
$[D_x^2 y]_{x=x_0} = 0$,
$[D_x^3 y]_{x=x_0} = 0$,
and $[D_x^4 y]_{x=x_0} < 0$, y_0 is a maximum.

(3) If $[D_x y]_{x=x_0} = 0$,
$[D_x^2 y]_{x=x_0} = 0$,
$[D_x^3 y]_{x=x_0} = 0$,
and $[D_x^4 y]_{x=x_0} > 0$, y_0 is a minimum.

47. The preceding investigation suggests the following method of finding the values of the variable corresponding to maximum or minimum values of the function. *Differentiate the function and find what values of* x *will make the first derivative zero.* This may, of course, be done by writing the derivative equal to zero, and solving the equation thus formed. *Substitute for* x, *in turn, in the second derivative, the values of* x *thus obtained, and note the signs of the results.* Those values of x which *make the second derivative positive* correspond to *minimum* values of

the function, and those that *make the second derivative negative*, to *maximum* values of the function. If any *make the second derivative zero*, they must be substituted for x in the *third derivative*, and the result interpreted by the method of Art. 46.

EXAMPLES.

Find what values of x give maximum and minimum values of the following functions: —

(1) $u = 2x^3 - 21x^2 + 36x - 20$.

$\qquad$ Ans. $x = 1$, max.; $x = 6$, min.

(2) $u = x^3 - 9x^2 + 15x - 3$.

$\qquad$ Ans. $x = 1$, max.; $x = 5$, min.

(3) $u = 3x^5 - 125x^3 + 2160x$.

$\qquad$ Ans. Max. when $x = -4$ or 3;
$\qquad$ min. when $x = -3$ or 4.

(4) Show that $\quad u = x^3 - 3x^2 + 6x + 7$

has neither a maximum or a minimum value; and that

$$u = x^5 - 5x^4 + 5x^3 - 1$$

is neither a maximum nor a minimum when $x = 0$.

(5) A person in a boat, three miles from the nearest point of the beach, wishes to reach, in the shortest possible time, a place

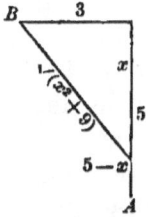

five miles from that point, along the shore. Supposing he can walk five miles an hour, but can row only four miles an hour, required the point of the beach he must pull for.

With the notation in the figure, the distance rowed is $\sqrt{(x^2+9)}$ miles, the distance walked is $5-x$ miles, and u, the whole time taken, is evidently

$$u = \frac{\sqrt{(x^2+9)}}{4} + \frac{5-x}{5} \qquad \text{hours,}$$

and x must have a value that will make u a minimum.

$$D_x u = \frac{x}{4\sqrt{(x^2+9)}} - \frac{1}{5},$$

$$D_x^2 u = \frac{9}{4(x^2+9)^{\frac{3}{2}}}.$$

Solving $\qquad \dfrac{x}{4\sqrt{(x^2+9)}} - \dfrac{1}{5} = 0,$

we get $\qquad x = \pm 4;$

but, on substituting these values of x in turn in the expression for $D_x u$, we see that $x=4$ is the only value which will make $D_x u = 0$, since we must take the positive value of $\sqrt{(x^2+9)}$, from the nature of the case, as it represents a distance traversed. Remembering this fact, we find

$$[D_x^2 u]_{x=4} = \frac{9}{500};$$

and u then is a minimum when $x=4$, and the landing-place must be one mile above the point of destination.

48. In problems concerning maxima and minima, the function u can often be most conveniently expressed in terms of two variables, x and y, which are themselves connected by some equation, so that either may be regarded as a function of the other. In this case, of course, u can, by elimination, be expressed in terms of either variable, and treated by the usual process. It is generally simpler, however, to differentiate u, regarding one of the variables, x, as the independent variable, and the other as a function of it, and then to substitute for $D_x y$

its value obtained from the given equation between x and y by the process suggested in Art. 29.

EXAMPLES.

(1) Required the maximum rectangle of given perimeter.
If a be the given perimeter, we have

$$2x + 2y = a; \qquad (1)$$

and the area $\quad u = xy. \qquad (2)$

Differentiate (1) with respect to x, and we have

$$2 + 2 D_x y = 0,$$

whence $\quad D_x y = -1; \qquad (3)$

$$D_x u = x D_x y + y = -x + y, \qquad \text{by (3),}$$

$$D_x^2 u = -1 + D_x y = -1 - 1 = -2, \qquad \text{by (3).}$$

$$D_x u = 0 \text{ if } x = y,$$

and $D_x^2 u$ is negative; therefore the required maximum rectangle is a square.

(2) Prove that of all circular sectors of given perimeter the greatest is that in which the arc is double the radius.

(3) A Norman window consists of a rectangle surmounted by a semicircle. Given the perimeter, required the height and breadth of the window when the quantity of light admitted is a maximum. *Ans.* Height and breadth must be equal.

49. After finding the values of x which make

$$D_x u = 0,$$

it is often possible to discriminate between those corresponding to maximum values of u and those corresponding to minimum values of u by outside considerations depending upon the nature of the problem, and so to avoid the labor of investigating the second derivative.

EXAMPLES.

(1) Prove that when the portion of a tangent to a circle intercepted between a pair of rectangular axes is a minimum it is equal to a diameter.

(2) Determine the greatest cylinder of revolution that can be inscribed in a given cone of revolution.

Ans. If b be the altitude of the cone and a the radius of its base, the volume of the required cylinder $= \dfrac{4}{27}\pi a^2 b$.

(3) Determine the cylinder of greatest convex surface that can be inscribed in the same cone.   *Ans.* Surface $= \dfrac{\pi b a}{2}$.

(4) Determine the cylinder of greatest convex surface that can be inscribed in a given sphere.   *Ans.* Altitude $= r\sqrt{(2)}$.

(5) Determine the greatest cone of revolution that can be inscribed in a given sphere.   *Ans.* Altitude $= \dfrac{4}{3}r$.

(6) Determine the cone of revolution of greatest convex surface that can be inscribed in a given sphere.

Ans. Altitude $= \dfrac{4}{3}r$.

Integration.

50. We have seen (Art. 12) that when a body moves according to any law, if v, t, and s are the velocity, time, and distance of the motion respectively, $v = D_t s$.

Suppose we have an expression for the velocity of a body in terms of the time during which it has been moving, and want to find the distance it has traversed. For example; the velocity of a falling body that has been falling t seconds is always gt, where g is constant at any given point of the earth's surface: required the distance fallen in t seconds.

This distance is evidently a function of t, for a change in the number of seconds a body falls changes the distance fallen.

Represent this function by s; then, as

$$v = D_t s,$$

we have $\qquad D_t s = gt;$

that is, *the distance is that function of t which has gt for its derivative;* and to solve the problem we have *to find the function when its derivative is given.*

51. Having given the equation $y = fx$ of a curve (rectangular coördinates), required the *area bounded by the curve, the axis of* X, *a fixed ordinate* y_0, *and any second ordinate* y.

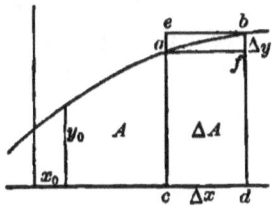

This area, A, is obviously a function of x, the abscissa corresponding to the second bounding ordinate y, for a change in x changes A. Let us see if we cannot find the value of $D_x A$. Increase x by Δx, and represent the corresponding increments of A and y by ΔA and Δy. From the figure, the area

$$acdf < acdb < ecdb;$$

but the area of the rectangle

$$acdf = y \Delta x,$$

the area of the rectangle

$$ecdb = (y + \Delta y) \Delta x,$$

and $\qquad acdb = \Delta A;$

hence $\qquad y \Delta x < \Delta A < (y + \Delta y) \Delta x;$

and we want $\underset{\Delta x \doteq 0}{\text{limit}} \left[\dfrac{\Delta A}{\Delta x} \right]$. Divide by Δx,

and
$$y < \frac{\Delta A}{\Delta x} < y + \Delta y.$$

That is, $\dfrac{\Delta A}{\Delta x}$ always lies between y and $y + \Delta y$; and as they approach the same limit, y, as $\Delta x \doteq 0$, $\underset{\Delta x \doteq 0}{\text{limit}} \left[\dfrac{\Delta A}{\Delta x} \right]$ must be y, and

we have
$$D_x A = y = fx;$$

and to solve the problem completely, we have to find a function from its derivative.

52. Having given the equation $y = fx$ of a curve (rectangular coördinates), required *the length of the arc between a fixed point* (x_0, y_0) *of the curve and any second point* (x, y).

This length is obviously a function of the position, and therefore

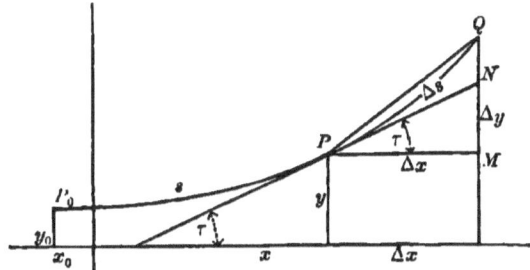

of the coördinates of the second point; and as the equation of the curve enables us to express y in terms of x, we can consider the length s a function of x. Let us see if we can find its derivative. Increase x by Δx and represent the corresponding increments of s and y by Δs and Δy respectively. We see from the figure that
$$PQ < \Delta s < PN + NQ,$$

PN being the tangent at P.
$$PQ = \sqrt{(\Delta x)^2 + (\Delta y)^2},$$

$$PN = \Delta x \cdot \sec \tau.$$

$$NQ = \Delta y - MN.$$

$$MN = \Delta x \cdot \tan \tau.$$

hence
$$NQ = \Delta y - \Delta x \tan \tau,$$
and we have

$$\sqrt{(\Delta x)^2 + (\Delta y)^2} < \Delta s < \Delta x \sec \tau + \Delta y - \Delta x \tan \tau.$$

Divide by Δx, —

$$\sqrt{1 + \left(\frac{\Delta y}{\Delta x}\right)^2} < \frac{\Delta s}{\Delta x} < \sec \tau + \frac{\Delta y}{\Delta x} - \tan \tau.$$

$$\lim_{\Delta x \doteq 0} \left[\sqrt{1 + \left(\frac{\Delta y}{\Delta x}\right)^2} \right] = \sqrt{1 + (D_x y)^2}$$

and
$$\lim_{\Delta x \doteq 0} \left[\sec \tau + \frac{\Delta y}{\Delta x} - \tan \tau \right] = \sec \tau + D_x y - \tan \tau.$$

But we know, Art. 27, [1], that

$$\tan \tau = D_x y;$$

and by trigonometry,

$$\sec^2 \tau = 1 + \tan^2 \tau = 1 + (D_x y)^2,$$

$$\sec \tau = \sqrt{1 + (D_x y)^2};$$

hence

$$\lim_{\Delta x \doteq 0} \left[\sec \tau + \frac{\Delta y}{\Delta x} - \tan \tau \right] = \sqrt{1 + (D_x y)^2} + D_x y - D_x y,$$

or
$$= \sqrt{1 + (D_x y)^2}.$$

As $\dfrac{\Delta s}{\Delta x}$ lies always between two quantities which have the same limit, $\sqrt{1 + (D_x y)^2}$, its limit must be $\sqrt{1 + (D_x y)^2}$, and we have

$$D_x s = \sqrt{1 + (D_x y)^2}.$$

$D_x y$ can be found from the given equation, and therefore

$\sqrt{1+(D_x y)^2}$ can be determined. We can then regard $D_x s$ as given, and again we are required to obtain a function from its derivative.

53. To find a function from its derivative is *to integrate*, and the function is called the *integral* of the given derivative. Thus the integral of $2x$ is $x^2 + C$, where C is any constant, for $D_x(x^2 + C)$ is $2x$. In other words, if y is a function of x, that function of x which has y for its derivative is called the *integral of* y *with respect to* x, and is indicated by $\int_x y$, the symbol $\int_x$ standing for the words *integral with respect to* x.

54. Since the derivative of a constant is zero, we may add any constant to a function without affecting the derivative of the function; so that if we know merely the value of the derivative, the function is not wholly determined, but may contain any *arbitrary*, i.e., undetermined, *constant* term. In special problems, there are usually sufficient additional data to enable us to determine this constant after effecting the integration.

55. Since *integration* is defined as *the inverse of differentiation*, we ought to be able to obtain a partial set of formulas for integrating by reversing the formulas we have already obtained for differentiating. Take the formulas —

$$D_x x = 1;$$

$$D_x ax = a;$$

$$D_x ay = a D_x y;$$

$$D_x x^n = n x^{n-1};$$

$$D_x(u + v + w + \&c.) = D_x u + D_x v + D_x w + \&c.;$$

and we get immediately —

$$\int_x 1 = x + C; \tag{1}$$

$$\int_x a = ax + C; \tag{2}$$

$$\int_{x} a D_{x} y = a y + C; \tag{3}$$

$$\int_{x} n x^{n-1} = x^{n} + C; \tag{4}$$

$$\int_{x} (D_{x} u + D_{x} v + D_{x} w + \&c.) = u + v + w + \&c. + C; \tag{5}$$

where C in each case is an arbitrary constant.

The forms of the last three can be modified with advantage.

In (3), call $\quad D_{x} y = u;$

then $\quad y = \int_{x} u,$

and (3) becomes $\quad \int_{x} a u = a \int_{x} u + C. \tag{6}$

By the aid of (6), (4) can be written,

$$n \int_{x} x^{n-1} = x^{n} + C.$$

Change n into $n+1$, and we get

$$(n+1) \int_{x} x^{n} = x^{n+1} + C,$$

or $\quad \int_{x} x^{n} = \dfrac{x^{n+1}}{n+1} + C, \tag{7}$

where C is any arbitrary constant, although, strictly speaking, different from the C just above.

In (5), let $\quad D_{x} u = y, \; D_{x} v = z, \; \&c.,$

then $\quad u = \int_{x} y, \; v = \int_{x} z, \; \&c.$

and $\quad \int_{x} (y + z + \&c.) = \int_{x} y + \int_{x} z + \&c. + C;$

or, *the integral of a sum of terms is the sum of the integrals of the terms.*

56. We can now solve the problem stated in Art. 50. The velocity of a falling body at the end of t seconds is gt feet, g

being a constant number; required the distance fallen in t seconds. We have seen that, if v, t, and s are the velocity, time, and distance respectively, $v = D_t s$;

hence $\qquad s = f_t v.$

Here $\qquad s = f_t gt + C;$

but by Art. 55, (6) and (7),

$$f_t gt = g f_t t = g \frac{t^2}{2} + C = \tfrac{1}{2} g t^2 + C;$$

and in this case we can readily determine C, for when the body has been falling no time, it has fallen no distance, so s must equal zero when $t = 0$, and we have

$$0 = \tfrac{1}{2} g (0)^2 + C = 0 + C, \text{ and } C = 0;$$

and our required result is $\quad s = \tfrac{1}{2} g t^2.$

57. Required the area intercepted by the curve $y^2 = 4x$, the axis of X, and the ordinate through the focus.

From the form of the equation we know that the curve is a parabola with its vertex at the origin and its focus at the point $(1,0)$. The initial ordinate in this case is evidently the tangent at the vertex.

If A is the required area, $D_x A = y,$ \hfill (Art. 51),

then $\qquad A = f_x y.$

$$y = 2\sqrt{x} = 2 x^{\tfrac{1}{2}};$$

hence $\qquad A = f_x 2 x^{\tfrac{1}{2}} = 2 f_x x^{\tfrac{1}{2}} = \dfrac{2 x^{\tfrac{3}{2}}}{\tfrac{3}{2}} + C = \dfrac{4}{3} x^{\tfrac{3}{2}} + C.$

A stands for the area terminated by the ordinate corresponding to any abscissa x.

It is obvious from the figure that if we make $x = 0$, the terminating ordinate y will coincide with the initial ordinate through the origin, and A will equal zero. So we can readily determine C,

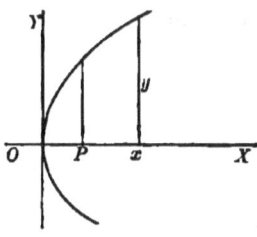

for we have $A = 0$ if $x = 0$;

so that $0 = \frac{4}{3} 0^{\frac{3}{2}} + C = C,$

and $A = \frac{4}{3} x^{\frac{3}{2}}.$

If $x = 1$, as it must in order that y may pass through the focus,

$$A = \frac{4}{3},$$ the required area.

Examples.

(1) Find the area bounded by the curve $x^2 = 4y$, the axis of X, and the ordinates corresponding to the abscissas 2 and 8.

Ans. 42.

(2) Prove that the area cut off from a parabola by a double ordinate is two-thirds of the circumscribing rectangle.

(3) Required the area intercepted between the curves $y^2 = 4ax$ and $x^2 = 4ay$. *Ans.* $\dfrac{16 a^2}{3}$.

(4) Find a formula for the area bounded by a curve $x = fy$, the axis of Y, and two lines parallel to the axis of abscissas.

Ans. $A = \int_y x + C$.

(5) Find a formula for the area intercepted by a curve $y = fx$, the axis of X, and two ordinates (oblique coördinates).

Ans. $A = \sin \omega \int_x y + C$, ω being the inclination of the axes.

(6) Prove that the segment of a parabola cut off by any chord is two-thirds of the circumscribing parallelogram.

58. Required the length of the portion of the line

$$4x - 3y + 2 = 0 \qquad (1)$$

between the points having the abscissas 1 and 4.

We have seen that $D_x s = \sqrt{1 + (D_x y)^2}$ \hfill Art. 52,

where s is the length of an arc;

hence $\qquad s = \int_x \sqrt{1 + (D_x y)^2}.$

From (1) we get $\qquad 4 - 3 D_x y = 0,$

$$D_x y = \tfrac{4}{3},$$

$$\sqrt{1 + (D_x y)^2} = \tfrac{5}{3};$$

and therefore $\qquad s = \int_x \tfrac{5}{3} = \tfrac{5}{3} x + C,$ \qquad where s stands for the length of the arc from the first point to any second point whose abscissa is x. If we make $x = 1$, the two points will coincide and s must equal 0; then $\qquad 0 = \tfrac{5}{3} + C,$

$$C = -\tfrac{5}{3},$$

and $\qquad s = \tfrac{5}{3}(x - 1).$

To get the required distance, x must equal 4 and we get $s = 5$.

Example.

Find the length of the portion of the line $Ax + By + C = 0$ between the points whose abscissas are x_0 and x_1.

$$\text{Ans. } \frac{\sqrt{(A^2 + B^2)}}{B}(x_1 - x_0).$$

CHAPTER IV.

TRANSCENDENTAL FUNCTIONS.

59. In order to complete our list of formulas for differentiating, we must consider the *transcendental* forms, $\log x$, a^x, $\sin x$, &c. Let us *differentiate* $\log x$.

By our fundamental method, we have

$$D_x \log x = \underset{\Delta x \doteq 0}{\text{limit}} \left[\frac{\log(x + \Delta x) - \log x}{\Delta x} \right],$$

$$\frac{\log(x + \Delta x) - \log x}{\Delta x} = \frac{1}{\Delta x} \log\left[\frac{x + \Delta x}{x}\right] = \frac{1}{\Delta x} \log\left[1 + \frac{\Delta x}{x}\right],$$

and $\qquad D_x \log x = \underset{\Delta x \doteq 0}{\text{limit}} \left[\frac{1}{\Delta x} \log\left(1 + \frac{\Delta x}{x}\right) \right].$

But as Δx approaches zero, $\log\left(1 + \frac{\Delta x}{x}\right)$ approaches $\log 1$, i.e., zero, and $\frac{1}{\Delta x}$ increases indefinitely; so that it is by no means easy to discover the limit of the product $\frac{1}{\Delta x} \log\left(1 + \frac{\Delta x}{x}\right)$.

This product can be thrown into a simpler form by introducing $\qquad m = \frac{x}{\Delta x} \qquad$ in place of Δx.

$\frac{1}{\Delta x} \log\left(1 + \frac{\Delta x}{x}\right)$ then becomes $\frac{m}{x} \log\left(1 + \frac{1}{m}\right)$, or $\frac{1}{x} \log\left(1 + \frac{1}{m}\right)^m$.

As Δx approaches zero, $\frac{x}{\Delta x}$ or m increases indefinitely, and

$$D_x \log x = \underset{m = \infty}{\text{limit}} \left[\frac{1}{x} \log\left(1 + \frac{1}{m}\right)^m \right],$$

and the value we have to investigate is the value approached by $\left(1+\dfrac{1}{m}\right)^m$ as its limit as m increases indefinitely, which we indicate by $\displaystyle\lim_{m=\infty}\left[1+\dfrac{1}{m}\right]^m$.

60. *Let us first suppose that* m *in its increase continues always a positive integer.* Then we can expand $\left(1+\dfrac{1}{m}\right)^m$ by the Binomial Theorem.

$$\left(1+\frac{1}{m}\right)^m = 1 + \frac{m}{1}\left(\frac{1}{m}\right) + \frac{m(m-1)}{1.2}\left(\frac{1}{m}\right)^2 + \frac{m(m-1)(m-2)}{1.2.3}\left(\frac{1}{m}\right)^3$$

$$+ \&c. \text{ to } m+1 \text{ terms}$$

$$= 1 + \frac{1}{1} + \frac{1-\dfrac{1}{m}}{1.2} + \frac{\left(1-\dfrac{1}{m}\right)\left(1-\dfrac{2}{m}\right)}{1.2.3}$$

$$+ \frac{\left(1-\dfrac{1}{m}\right)\left(1-\dfrac{2}{m}\right)\left(1-\dfrac{3}{m}\right)}{1.2.3.4} + \ldots \text{ to } m+1 \text{ terms}.$$

Now, as m increases indefinitely, each of the first n terms of the series, n being any fixed number, approaches as its limit the corresponding term of the series

$$1 + \frac{1}{1} + \frac{1}{1.2} + \frac{1}{1.2.3} + \frac{1}{1.2.3.4} + \ldots;$$

so that we have reason to suppose that there is some simple relation between this latter series and our required limit.

61. To investigate this question we shall divide the first series into two parts. The first part, consisting of the first $n+1$ terms, where n is any fixed whole number less than m, we shall represent by S; the second part, consisting of the remaining $m-n$ terms, we shall call R.

CHAP. IV.] TRANSCENDENTAL FUNCTIONS.

Then
$$\left(1+\frac{1}{m}\right)^m = S + R.$$

$$S = 1 + \frac{1}{1} + \frac{1-\frac{1}{m}}{1.2} + \frac{\left(1-\frac{1}{m}\right)\left(1-\frac{2}{m}\right)}{1.2.3} + \ldots$$

$$+ \frac{\left(1-\frac{1}{m}\right)\left(1-\frac{2}{m}\right)\cdots\left(1-\frac{n-1}{m}\right)}{1.2.3\ldots n}.$$

As n is a fixed number, we have

$$\lim_{m=\infty} [S] = 1 + \frac{1}{1} + \frac{1}{1.2} + \frac{1}{1.2.3} + \ldots + \frac{1}{1.2.3\ldots n}.$$

$$R = \frac{\left(1-\frac{1}{m}\right)\left(1-\frac{2}{m}\right)\cdots\left(1-\frac{n-1}{m}\right)}{1.2.3\ldots n} \left[\frac{1-\frac{n}{m}}{n+1} + \frac{\left(1-\frac{n}{m}\right)\left(1-\frac{n+1}{m}\right)}{(n+1)(n+2)} \right.$$

$$\left. + \ldots + \frac{\left(1-\frac{n}{m}\right)\left(1-\frac{n+1}{m}\right)\left(1-\frac{n+2}{m}\right)\cdots\left(1-\frac{m-1}{m}\right)}{(n+1)(n+2)(n+3)\ldots m} \right].$$

Since n is less than m, each numerator in the value of R is positive and less than 1, and

$$R < \frac{1}{1.2.3\ldots n}\left[\frac{1}{n+1} + \frac{1}{(n+1)^2} + \frac{1}{(n+1)^3} + \ldots + \frac{1}{(n+1)^{m-n}} \right].$$

The sum of the decreasing geometrical series,

$$\frac{1}{n+1} + \frac{1}{(n+1)^2} + \frac{1}{(n+1)^3} + \ldots,$$

is by algebra less than $\frac{1}{n}$;

therefore
$$R < \frac{1}{n(1.2.3\ldots n)},$$

and
$$\lim_{m=\infty} [R] < \frac{1}{n(1.2.3.....n)};$$

and we have at last,

$$\lim_{m=\infty} \left(1+\frac{1}{m}\right)^m = \lim_{m=\infty} [S] + \lim_{m=\infty} [R]$$

$$= 1 + \frac{1}{1} + \frac{1}{1.2} + \frac{1}{1.2.3} + \frac{1}{1.2.3.4} + \cdots + \frac{1}{1.2.3.....n}$$

$$+ \text{something less than } \frac{1}{n(1.2.3......n)},$$

n *being any positive whole number.*

Thus we obtain the relation that the difference between our required value and the sum of the first $n+1$ terms of the series

$$1 + \frac{1}{1} + \frac{1}{1.2} + \frac{1}{1.2.3} + \cdots$$

is less than $\frac{1}{n(1.2.3.......n)}$.

The greater the value of n the less the value of $\frac{1}{n.1.2.3....n}$; and by taking a value of n sufficiently great, we may make this difference as small as we please.

Consequently, by Art. 7, our required value is the limit approached by the sum of the first n *terms of the series*

$$1 + \frac{1}{1} + \frac{1}{1.2} + \frac{1}{1.2.3} + \cdots$$

as n *is indefinitely increased,* or what is ordinarily called the *sum* of this series.

62. The series $1 + \frac{1}{1} + \frac{1}{1.2} + \frac{1}{1.2.3} + \cdots$ plays a very important part in the theory of logarithms. It is generally represented by the letter e, and is taken as the base of the natural system of logarithms. Its numerical value can be readily computed to any required number of decimal places, since each term of the

series may be obtained by dividing the preceding one by the number of the term minus one. Carrying the approximation to six decimal places, we have

```
1.
1.
0.5
0.166666
0.041666
0.008333
0.001388
0.000198 4
0.000024 8
0.000002 7
0.000000 2
0.000000 0
```

The error in the approximation is less than one-eleventh of the last term we have used, and therefore cannot affect our sixth decimal place.

$e = 2.718281+$, correct to six decimal places.

63. *Let us now remove from* m *the restriction we placed upon it when we supposed it to have none but positive integral values, and suppose it to increase passing through* all *positive values.* Let μ represent at any instant the integer next below m, then $\mu+1$ will be the integer next above m, and as m increases it will always be between μ and $\mu+1$, unless it happens to coincide with $\mu+1$, as it sometimes will. We have, then, in general,

$$\mu < m < \mu+1.$$

Then $\qquad \left(1+\dfrac{1}{\mu+1}\right)^{\mu} < \left(1+\dfrac{1}{m}\right)^{m} < \left(1+\dfrac{1}{\mu}\right)^{\mu+1};$

$\therefore \displaystyle\lim_{m=\infty} \left(1+\dfrac{1}{m}\right)^{m}$ must lie between $\displaystyle\lim_{\mu=\infty}\left(1+\dfrac{1}{\mu+1}\right)^{\mu}$ and $\displaystyle\lim_{\mu=\infty}\left(1+\dfrac{1}{\mu}\right)^{\mu+1},$

$\left(1+\dfrac{1}{\mu+1}\right)^{\mu} = \dfrac{\left(1+\dfrac{1}{\mu+1}\right)^{\mu+1}}{1+\dfrac{1}{\mu+1}}$, and $\displaystyle\lim_{\mu=\infty} \dfrac{\left(1+\dfrac{1}{\mu+1}\right)^{\mu+1}}{1+\dfrac{1}{\mu+1}} = \dfrac{e}{1} = e.$

$$\left(1+\frac{1}{\mu}\right)^{\mu+1} = \left(1+\frac{1}{\mu}\right)^{\mu}\left(1+\frac{1}{\mu}\right),$$

and $\quad \lim\limits_{\mu=\infty} \left(1+\frac{1}{\mu}\right)^{\mu}\left(1+\frac{1}{\mu}\right) = e \times 1 = e;$

hence $\quad \lim\limits_{m=\infty} \left(1+\frac{1}{m}\right)^{m} = e.$

Again: let m be negative, and represent it by $-r$,

then $\left(1+\dfrac{1}{m}\right)^{m} = \left(1-\dfrac{1}{r}\right)^{-r} = \left(\dfrac{r-1}{r}\right)^{-r} = \left(\dfrac{r}{r-1}\right)^{r} = \left(1+\dfrac{1}{r-1}\right)^{r}$

$$= \left(1+\frac{1}{r-1}\right)^{r-1}\left(1+\frac{1}{r-1}\right),$$

and $\quad \lim\limits_{m=\infty} \left(1+\dfrac{1}{m}\right)^{m}$

$= \lim\limits_{(r-1)=\infty} \left(1+\dfrac{1}{r-1}\right)^{r-1}\left(1+\dfrac{1}{r-1}\right) = e \times 1 = e.$

We see, then, that always

$$\lim\limits_{m=\infty} \left(1+\frac{1}{m}\right)^{m} = e = 2.718281 + \cdot$$

64. In Art. 59 we found that

$$D_z \log x = \lim\limits_{m=\infty} \left[\frac{1}{x} \log\left(1+\frac{1}{m}\right)^{m}\right].$$

We have, then, $\quad D_z \log x = \dfrac{1}{x} \log e.$

If by $\log x$ we mean, as we shall always mean hereafter, *natural logarithm* of x, $\log e$ will equal 1, and

$$D_x \log x = \frac{1}{x}. \qquad [1]$$

If $y = fx$,
$$D_x \log y = \frac{D_x y}{y}.$$

Exponential Functions.

65. *Required* $D_x a^x$, *a being any constant.*

Let
$$u = a^x$$
and take the log of each member,
$$\log u = x \log a.$$

Take D_x of both members,
$$\frac{D_x u}{u} = \log a;$$
$$D_x u = u \log a,$$
$$D_x a^x = a^x \log a. \qquad [1]$$

If $a = e$; since $\log e = 1$,
we have
$$D_x e^x = e^x. \qquad [2]$$

Of course,
$$D_x a^y = a^y \log a \, D_x y,$$
and
$$D_x e^y = e^y D_x y.$$

EXAMPLES.

Find $D_x u$ in each of the following cases:—

(1) $u = e^x(1-x^3)$.   Ans. $D_x u = e^x(1 - 3x^2 - x^3)$.

(2) $u = \dfrac{e^x - e^{-x}}{e^x + e^{-x}}$.   Ans. $D_x u = \dfrac{4}{(e^x + e^{-x})^2}$.

(3) $u = \log(e^x + e^{-x})$. Ans. $D_x u = \dfrac{e^x - e^{-x}}{e^x + e^{-x}}$.

(4) $u = \dfrac{x}{e^x - 1}$. Ans. $D_x u = \dfrac{e^x(1-x) - 1}{(e^x - 1)^2}$.

(5) $u = \log \dfrac{\sqrt{(1+x)} + \sqrt{(1-x)}}{\sqrt{(1+x)} - \sqrt{(1-x)}}$. Ans. $D_x u = -\dfrac{1}{x\sqrt{(1-x^2)}}$.

(6) $u = \log(\log x)$. Ans. $D_x u = \dfrac{1}{x \log x}$.

(7) $u = \log \dfrac{x}{x + \sqrt{(1+x^2)}}$. Ans. $D_x u = \dfrac{1}{x} - \dfrac{1}{\sqrt{(1+x^2)}}$.

(8) $u = x^x$. Ans. $D_x u = x^x(\log x + 1)$.

Suggestion. Take the log of each member before differentiating.

(9) $u = x^{\frac{1}{x}}$. Ans. $D_x u = \dfrac{x^{\frac{1}{x}}(1 - \log x)}{x^2}$.

(10) $u = e^{e^x}$. Ans. $D_x u = e^{e^x} e^x$.

(11) $u = e^{x^x}$. Ans. $D_x u = e^{x^x} x^x (1 + \log x)$.

(12) $u = x^{e^x}$. Ans. $D_x u = x^{e^x} e^x \dfrac{1 + x \log x}{x}$.

Trigonometric Functions.

66. In higher mathematics *an angle is represented numerically,* not by the number of degrees it contains but *by the ratio of the length of its arc to the length of the radius with which the arc is described.*

Thus the angle θ is said to be equal to $\dfrac{\text{arc } \theta}{r}$. If the arc is

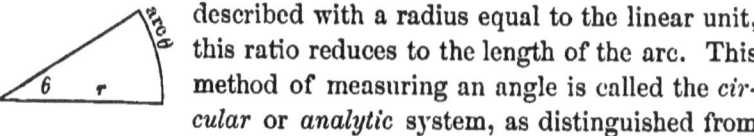

described with a radius equal to the linear unit, this ratio reduces to the length of the arc. This method of measuring an angle is called the *circular* or *analytic* system, as distinguished from the ordinary *degree* or *gradual* system.

The value of 360° in circular measure is obviously $\frac{2\pi r}{r}$ or 2π, and of 1° is $\frac{2\pi}{360}$ or $\frac{\pi}{180}$. Hence, *to reduce from gradual to circular measure, it is only necessary to multiply the given number of degrees by* $\frac{\pi}{180}$.

The circular unit is evidently the angle which has its arc equal to the radius, and its value in degrees is easily found. Let x represent the required value in degrees; then

$$\frac{x°}{360°} = \frac{r}{2\pi r} \text{ and } x = \frac{180°}{\pi}.$$

Hence, *to reduce from circular to gradual measure, we have only to multiply the circular value by* $\frac{180}{\pi}$.

67. *Required* $D_x \sin x$.

By our usual method, we have

$$D_x \sin x = \lim_{\Delta x \doteq 0} \left[\frac{\sin(x + \Delta x) - \sin x}{\Delta x} \right],$$

$$\frac{\sin(x + \Delta x) - \sin x}{\Delta x} = \frac{\sin x \cos \Delta x + \cos x \sin \Delta x - \sin x}{\Delta x}$$

$$= \frac{\cos x \sin \Delta x - \sin x (1 - \cos \Delta x)}{\Delta x}.$$

$$D_x \sin x = \lim_{\Delta x \doteq 0} \left[\cos x \frac{\sin \Delta x}{\Delta x} - \sin x \frac{1 - \cos \Delta x}{\Delta x} \right]$$

$$= \cos x \lim_{\Delta x \doteq 0} \left[\frac{\sin \Delta x}{\Delta x} \right] - \sin x \lim_{\Delta x \doteq 0} \left[\frac{1 - \cos \Delta x}{\Delta x} \right].$$

But as $\Delta x \doteq 0$, $\sin \Delta x \doteq 0$ and $\cos \Delta x \doteq 1$, so both of our limits, in their present form, are indeterminate, and require special investigation.

68. Suppose an arc described from the vertex of the angle $\varDelta x$, 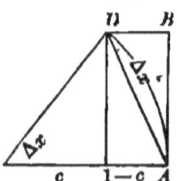 with a radius equal to unity, then this arc measures the angle, and is equal to $\varDelta x$, and the lengths of the lines, marked s and c in the figure, are the $\sin \varDelta x$ and $\cos \varDelta x$, respectively.

We wish to find $\underset{\varDelta x \doteq 0}{\text{limit}} \left[\dfrac{s}{\varDelta x} \right]$ and $\underset{\varDelta x \doteq 0}{\text{limit}} \left[\dfrac{1-c}{\varDelta x} \right]$.

$$\text{arc } \varDelta x < AB + BD$$

by geometry (*vide* "Chauvenet's Geometry," Book V. Prop. xii.).

We have then $\quad AD < \varDelta x < AB + BD$;

or, since $\quad s < AD$, and $AB + BD = s + 1 - c$,

$$s < \varDelta x < s + 1 - c.$$

But
$$s^2 + c^2 = 1,$$

$$1 - c^2 = s^2,$$

$$1 - c = \frac{s^2}{1+c},$$

and
$$s + 1 - c = \frac{s(1+c) + s^2}{1+c};$$

hence
$$s < \varDelta x < \frac{s(1+c) + s^2}{1+c};$$

$$\therefore \frac{s}{s} > \frac{s}{\varDelta x} > \frac{s(1+c)}{s(1+c) + s^2},$$

or
$$1 > \frac{s}{\varDelta x} > \frac{1+c}{1+c+s},$$

and $\underset{\varDelta x \doteq 0}{\text{limit}} \left[\dfrac{s}{\varDelta x} \right]$ must be between 1 and $\underset{\varDelta x \doteq 0}{\text{limit}} \left[\dfrac{1+c}{1+c+s} \right]$; but since, as $\varDelta x \doteq 0$, $s \doteq 0$, and $c \doteq 1$,

CHAP. IV.] TRANSCENDENTAL FUNCTIONS. 59

$$\underset{\Delta x \doteq 0}{\text{limit}} \left[\frac{1+c}{1+c+s} \right] = \frac{2}{2} = 1;$$

therefore $\underset{\Delta x \doteq 0}{\text{limit}} \left[\frac{\sin \Delta x}{\Delta x} \right] = 1.$

In like manner, $\dfrac{1-c}{s} > \dfrac{1-c}{\Delta x} > \dfrac{(1-c)(1+c)}{s(1+c)+s^2}$

or $\dfrac{s^2}{s(1+c)} > \dfrac{1-c}{\Delta x} > \dfrac{s^2}{s(1+c)+s^2},$

$\dfrac{s}{1+c} > \dfrac{1-c}{\Delta x} > \dfrac{s}{1+c+s};$

$\underset{\Delta x \doteq 0}{\text{limit}} \left[\dfrac{1-c}{\Delta x} \right]$ lies between $\underset{\Delta x \doteq 0}{\text{limit}} \left[\dfrac{s}{1+c} \right]$, or 0, and $\underset{\Delta x \doteq 0}{\text{limit}} \left[\dfrac{s}{1+c+s} \right]$, or 0;

therefore $\underset{\Delta x \doteq 0}{\text{limit}} \left[\dfrac{1-\cos \Delta x}{\Delta x} \right] = 0.$

69. Substituting these values in Art. 67, we have

$$D_x \sin x = \cos x.$$

EXAMPLES.

(1) Prove $\quad D_x \cos x = - \sin x.$

(2) Prove $\quad D_x \tan x = \sec^2 x,$

$D_x \operatorname{ctn} x = - \csc^2 x,$

$D_x \sec x = \tan x \sec x,$

$D_x \csc x = - \operatorname{ctn} x \csc x,$

from the relations $\quad \tan x = \dfrac{\sin x}{\cos x},$

$$\operatorname{ctn} x = \frac{1}{\tan x},$$

$$\sec x = \frac{1}{\cos x},$$

$$\csc x = \frac{1}{\sin x}.$$

(3) Given $\operatorname{vers} x = 1 - \cos x,$

prove $D_x \operatorname{vers} x = \sin x.$

4) Prove $D_x \log \sin x = \operatorname{ctn} x;$

$D_x \log \cos x = -\tan x;$

$D_x \log \tan x = \sec x \csc x;$

$D_x \log \operatorname{ctn} x = -\sec x \csc x;$

$D_x \log \sec x = \tan x;$

$D_x \log \csc x = -\operatorname{ctn} x.$

Anti-Trigonometric Functions.

70. In trigonometry, the *angle which has a sine equal to* x is called *the inverse sine or the anti-sine of* x, and is denoted by the symbol $\sin^{-1}$. Hence $\sin^{-1} x$ means the angle which has x for its sine, and is to be read *anti-sine* of x.

In the same way we speak of *anti-cosine, anti-tangent,* &c.

71. *To differentiate* $\sin^{-1} x$.

Let $y = \sin^{-1} x;$ then $x = \sin y.$

Differentiate both members with respect to x.

$$1 = \cos y \, D_x y;$$

$$D_x y = \frac{1}{\cos y}.$$

It remains to express $\cos y$ in terms of x.

$$\sin y = x,$$
$$\cos^2 y = 1 - x^2,$$
$$\cos y = \sqrt{(1-x^2)};$$

hence
$$D_x \sin^{-1} x = \frac{1}{\sqrt{(1-x^2)}}.$$

EXAMPLES.

(1) Prove $\quad D_x \cos^{-1} x = -\dfrac{1}{\sqrt{(1-x^2)}}.$

(2) $\quad D_x \tan^{-1} x = \dfrac{1}{1+x^2}.$

(3) $\quad D_x \operatorname{ctn}^{-1} x = -\dfrac{1}{1+x^2}.$

(4) $\quad D_x \sec^{-1} x = \dfrac{1}{x\sqrt{(x^2-1)}}.$

(5) $\quad D_x \csc^{-1} x = -\dfrac{1}{x\sqrt{(x^2-1)}}.$

(6) $\quad D_x \operatorname{vers}^{-1} x = \dfrac{1}{\sqrt{(2x-x^2)}}.$

72. The *anti-*, or *inverse*, *notation* is not confined to trigonometric functions. The number which has x for its logarithm is called the *anti-logarithm* of x, and is denoted by $\log^{-1} x$; and, in general, *if* x *is any function of* y, y *may be called the corresponding anti-function of* x, and the relation of y to x will be indicated by the same functional symbol as that which expresses

the dependence of x upon y, except that it will be affected with a negative exponent, which, however, must not be confounded with a negative exponent in the algebraic sense. Thus, if $x=fy$, we may write $y=f^{-1}x$.

Any anti-function can be readily differentiated, if the direct function can be differentiated, and by the method we have employed in the case of the anti-trigonometric functions above.

Let
$$y = f^{-1}x,$$

then
$$x = fy;$$

differentiate, and
$$1 = D_y fy \cdot D_x y.$$

$$D_x y = \frac{1}{D_y fy}$$

or
$$D_x f^{-1}x = \frac{1}{D_y fy},$$

and it is only necessary to replace y in this result by its value in terms of x.

73. Since, in the formula above,

$$fy = x,$$

we have
$$D_x y = \frac{1}{D_y x};$$

a result so important that it is worth while to establish it by more elementary considerations.

Suppose x and y connected by any relation, so that either may be regarded as a function of the other. Let Δx and Δy be corresponding increments of x and y. Then Δx may be regarded as having produced Δy, or as having been produced by Δy, according as we regard x or y as the independent variable; and on either hypothesis they will approach zero together.

By definition,
$$D_x y = \lim_{\Delta x \doteq 0} \left[\frac{\Delta y}{\Delta x} \right],$$

CHAP. IV.] TRANSCENDENTAL FUNCTIONS. 63

and
$$D_y x = \underset{\Delta x \doteq 0}{\text{limit}} \left[\frac{\Delta x}{\Delta y} \right].$$

$$\frac{\Delta y}{\Delta x} = \frac{1}{\frac{\Delta x}{\Delta y}};$$

and
$$\underset{\Delta x \doteq 0}{\text{limit}} \left[\frac{\Delta y}{\Delta x} \right] = \frac{1}{\underset{\Delta x \doteq 0}{\text{limit}} \left[\frac{\Delta x}{\Delta y} \right]} = \frac{1}{\underset{\Delta y \doteq 0}{\text{limit}} \left[\frac{\Delta x}{\Delta y} \right]},$$

since Δx and Δy approach 0 together.

Therefore
$$D_x y = \frac{1}{D_y x}.$$

EXAMPLES.

Find $D_x u$ in the following cases: —

(1) $u = \sin^2 x.$ Ans. $D_x u = 2 \sin x \cos x.$

(2) $u = \cos mx.$ Ans. $D_x u = - m \sin mx.$

(3) $u = x e^{\cos x}.$ Ans. $D_x u = e^{\cos x}(1 - x \sin x).$

(4) $u = \cos(\sin x).$ Ans. $D_x u = - \cos x \sin(\sin x).$

(5) $u = \sin(\log x).$ Ans. $D_x u = \frac{1}{x} \cos(\log x).$

(6) $u = \dfrac{\tan^3 x}{3} - \tan x + x.$ Ans. $D_x u = \tan^4 x$

(7) $u = (a^2 + x^2) \tan^{-1} \dfrac{x}{a}.$ Ans. $D_x u = 2x \tan^{-1} \dfrac{x}{a} + a.$

(8) $u = x \sin^{-1} x.$ Ans. $D_x u = \sin^{-1} x + \dfrac{x}{\sqrt{(1-x^2)}}.$

(9) $u = \sin^{-1} \dfrac{x+1}{\sqrt{(2)}}.$ Ans. $D_x u = \dfrac{1}{\sqrt{(1-2x-x^2)}}.$

(10) $u = \tan^{-1} \dfrac{x}{\sqrt{(1-x^2)}}$.   Ans. $D_x u = \dfrac{1}{\sqrt{(1-x^2)}}$.

(11) $u = \sec^{-1} \dfrac{a}{\sqrt{(a^2-x^2)}}$.   Ans. $D_x u = \dfrac{1}{\sqrt{(a^2-x^2)}}$.

(12) $u = \sin^{-1} \sqrt{(\sin x)}$.   Ans. $D_x u = \tfrac{1}{2}\sqrt{(1+\csc x)}$.

(13) $u = \tan^{-1} \dfrac{2x}{1-x^2}$.   Ans. $D_x u = \dfrac{2}{1+x^2}$.

(14) $u = \tan^{-1} \sqrt{\left(\dfrac{1-\cos x}{1+\cos x}\right)}$.   Ans. $D_x u = \dfrac{1}{2}$.

CHAPTER V.

INTEGRATION.

74. We are now able to extend materially our list of *formulas for direct integration* (Art. 55), one of which may be obtained from each of the derivative formulas in our last chapter. The following set contains the most important of these: —

$D_x \log x = \dfrac{1}{x}$ gives $\displaystyle\int_x \dfrac{1}{x} = \log x.$

$D_x a^x = a^x \log a$ " $\displaystyle\int_x a^x \log a = a^x.$

$D_x e^x = e^x$ " $\displaystyle\int_x e^x = e^x.$

$D_x \sin x = \cos x$ " $\displaystyle\int_x \cos x = \sin x.$

$D_x \cos x = -\sin x$ " $\displaystyle\int_x (-\sin x) = \cos x.$

$D_x \log \sin x = \operatorname{ctn} x$ " $\displaystyle\int_x \operatorname{ctn} x = \log \sin x.$

$D_x \log \cos x = -\tan x$ " $\displaystyle\int_x (-\tan x) = \log \cos x.$

$D_x \sin^{-1} x = \dfrac{1}{\sqrt{(1-x^2)}}$ " $\displaystyle\int_x \dfrac{1}{\sqrt{(1-x^2)}} = \sin^{-1} x.$

$D_x \tan^{-1} x = \dfrac{1}{1+x^2}$ " $\displaystyle\int_x \dfrac{1}{1+x^2} = \tan^{-1} x.$

$D_x \operatorname{vers}^{-1} x = \dfrac{1}{\sqrt{(2x-x^2)}}$ " $\displaystyle\int_x \dfrac{1}{\sqrt{(2x-x^2)}} = \operatorname{vers}^{-1} x.$

The second, fifth, and seventh in the second group can be written in the more convenient forms,

$$\int_x a^x = \frac{a^x}{\log a};$$

$$\int_x \sin x = -\cos x;$$

$$\int_x \tan x = -\log \cos x.$$

75. When the expression to be integrated does not come under any of the forms in the preceding list, *it can often be prepared for integration by a suitable change of variable*, the new variable, of course, being a function of the old. This method is called *integration by substitution*, and is based upon a formula easily deduced from $\quad D_x(Fy) = D_y Fy \cdot D_x y$;

which gives immediately

$$Fy = \int_x (D_y Fy \cdot D_x y).$$

Let $\quad u = D_y Fy,$

then $\quad Fy = \int_y u,$

and we have $\quad \int_y u = \int_x (u D_x y);$

or, interchanging x and y,

$$\int_x u = \int_y (u D_y x). \qquad [1]$$

For example, required $\quad \int_x (a + bx)^n.$

Let $\quad z = a + bx,$

and then $\quad \int_x (a + bx)^n = \int_x z^n = \int_z (z^n \cdot D_z x), \qquad$ by [1];

but $\quad x = \frac{z}{b} - \frac{a}{b},$

$$D_z x = \frac{1}{b};$$

hence $\quad \int_x (a + bx)^n = \frac{1}{b} \int_z z^n = \frac{1}{b} \frac{z^{n+1}}{n+1}.$

Substituting for z its value, we have

$$\int_x (a+bx)^n = \frac{1}{b}\frac{(a+bx)^{n+1}}{n+1}.$$

EXAMPLE.

Find $\int_x \frac{1}{a+bx}$. Ans. $\frac{1}{b}\log(a+bx)$.

76. If fx represents a function that can be integrated, $f(a+bx)$ can always be integrated; for, if

$$z = a + bx,$$

then
$$D_x x = \frac{1}{b}$$

and
$$\int_x f(a+bx) = \int_x fz = \int_z fz D_x x = \frac{1}{b}\int_z fz.$$

EXAMPLES.

Find

(1) $\int_x \sin ax$. Ans. $-\frac{1}{a}\cos ax$.

(2) $\int_x \cos ax$. Ans. $\frac{1}{a}\sin ax$.

(3) $\int_x \tan ax$.

(4) $\int_x \operatorname{ctn} ax$.

77. Required $\int_x \frac{1}{\sqrt{(a^2-x^2)}}$.

$$\int_x \frac{1}{\sqrt{(a^2-x^2)}} = \frac{1}{a}\int_x \frac{1}{\sqrt{\left[1-\left(\frac{x}{a}\right)^2\right]}}.$$

Let
$$z = \frac{x}{a},$$

then
$$x = az,$$
$$D_x x = a,$$

$$\frac{1}{a}\int_z \frac{1}{\sqrt{\left[1-\left(\frac{x}{a}\right)^2\right]}} = \frac{1}{a}\int_z \frac{1}{\sqrt{(1-z^2)}} = \frac{1}{a}\int_z \frac{1}{\sqrt{(1-z^2)}} D_z x$$

$$= \int_z \frac{1}{\sqrt{(1-z^2)}} = \sin^{-1} z = \sin^{-1} \frac{x}{a}.$$

EXAMPLES.

Find

(1) $\int_x \dfrac{1}{a^2+x^2}.$ \hfill Ans. $\dfrac{1}{a}\tan^{-1}\dfrac{x}{a}.$

(2) $\int_x \dfrac{1}{\sqrt{(2ax-x^2)}}.$ \hfill Ans. $\text{vers}^{-1}\dfrac{x}{a}.$

78. Required $\int_x \dfrac{1}{\sqrt{(x^2+a^2)}}.$

Let $\qquad z = x + \sqrt{(x^2+a^2)}$;

then $\qquad z - x = \sqrt{(x^2+a^2)},$

$$z^2 - 2zx + x^2 = x^2 + a^2,$$

$$2zx = z^2 - a^2,$$

$$x = \frac{z^2 - a^2}{2z},$$

$$\sqrt{(x^2+a^2)} = z - x = z - \frac{z^2-a^2}{2z} = \frac{z^2+a^2}{2z},$$

$$D_z x = \frac{z^2+a^2}{2z^2}.$$

$$\int_x \frac{1}{\sqrt{(x^2+a^2)}} = \int_x \frac{2z}{z^2+a^2} = \int_z \frac{2z}{z^2+a^2} D_z x$$

$$= \int_z \frac{2z}{z^2+a^2} \cdot \frac{z^2+a^2}{2z^2} = \int_z \frac{1}{z} = \log z = \log(x + \sqrt{x^2+a^2}).$$

EXAMPLE.

Find $\int_x \dfrac{1}{\sqrt{(x^2-a^2)}}.$ \hfill Ans. $\log(x + \sqrt{x^2-a^2}).$

79. *When the expression to be integrated can be factored*, the required integral can often be obtained by the use of a formula deduced from

$$D_x(uv) = uD_xv + vD_xu,$$

which gives

$$uv = \int_x uD_xv + \int_x vD_xu$$

or

$$\int_x uD_xv = uv - \int_x vD_xu. \qquad [1]$$

This method is called *integrating by parts*.

(a) For example, required $\int_x \log x$.

$\log x$ can be regarded as the product of $\log x$ by 1.

Call $\qquad \log x = u$ and $1 = D_xv,$

then $\qquad D_xu = \dfrac{1}{x},$

$$v = x;$$

and we have

$$\int_x \log x = \int_x 1 \log x = \int_x uD_xv = uv - \int_x vD_xu$$

$$= x\log x - \int_x \dfrac{x}{x} = x\log x - x.$$

EXAMPLE.

Find $\int_x x\log x$.

Suggestion: Let $\log x = u$ and $x = D_xv.$

$$\text{Ans. } \tfrac{1}{2}x^2\left(\log x - \tfrac{1}{2}\right)$$

80. Required $\int_x \sin^2 x$.

Let $\qquad u = \sin x$ and $D_xv = \sin x,$

then $\qquad D_xu = \cos x,$

$$v = -\cos x,$$

$$\int_x \sin^2 x = -\sin x \cos x + \int_x \cos^2 x;$$

but
$$\cos^2 x = 1 - \sin^2 x,$$

so
$$\int_x \cos^2 x = \int_x 1 - \int_x \sin^2 x = x - \int_x \sin^2 x$$

and
$$\int_x \sin^2 x = x - \sin x \cos x - \int_x \sin^2 x.$$

$$2\int_x \sin^2 x = x - \sin x \cos x.$$

$$\int_x \sin^2 x = \tfrac{1}{2}(x - \sin x \cos x).$$

EXAMPLES.

(1) Find $\int_x \cos^2 x$. *Ans.* $\dfrac{1}{2}(x + \sin x \cos x)$.

(2) $\int_x \sin x \cos x$. *Ans.* $\dfrac{\sin^2 x}{2}$.

81. *Very often both methods described above are required in the same integration.*

(a) Required $\int_x \sin^{-1} x$.

Let
$$\sin^{-1} x = y,$$

then
$$x = \sin y;$$

$$D_y x = \cos y,$$

$$\int_x \sin^{-1} x = \int_x y = \int_y y \cos y.$$

Let $u = y$ and $D_y v = \cos y;$

then
$$D_y u = 1,$$

$$v = \sin y,$$

and
$$\int_y y \cos y = y \sin y - \int_y \sin y = y \sin y + \cos y = x \sin^{-1} x + \sqrt{(1-x^2)}.$$

Any inverse or anti-function can be integrated by this method if the direct function is integrable.

(b) Thus, $\int_x f^{-1} x = \int_x y = \int_y y D_y fy = yfy - \int_y fy$

where
$$y = f^{-1} x.$$

EXAMPLES.

(1) Find $\int_x \cos^{-1} x$.   Ans. $x \cos^{-1} x - \sqrt{(1-x^2)}$.

(2) $\int_x \tan^{-1} x$.   Ans. $x \tan^{-1} x - \frac{1}{2} \log(1+x^2)$.

(3) $\int_x \mathrm{vers}^{-1} x$.   Ans. $(x-1) \mathrm{vers}^{-1} x + \sqrt{(2x-x^2)}$.

82. Sometimes an *algebraic transformation*, either alone or in combination with the preceding methods, is useful.

(a) Required $\int_x \dfrac{1}{x^2 - a^2}$.

$$\frac{1}{x^2-a^2} = \frac{1}{2a}\left(\frac{1}{x-a} - \frac{1}{x+a}\right),$$

and, by Art. 75 (Ex.),

$$\int_x \frac{1}{x^2-a^2} = \frac{1}{2a}[\log(x-a) - \log(x+a)] = \frac{1}{2a}\log\frac{x-a}{x+a}.$$

(b) Required $\int_x \sqrt{\left(\dfrac{1+x}{1-x}\right)}$.

$$\sqrt{\left(\frac{1+x}{1-x}\right)} = \frac{1+x}{\sqrt{(1-x^2)}} = \frac{1}{\sqrt{(1-x^2)}} + \frac{x}{\sqrt{(1-x^2)}},$$

$$\int_x \frac{1}{\sqrt{(1-x^2)}} = \sin^{-1} x.$$

$\int_x \dfrac{x}{\sqrt{(1-x^2)}}$ can be readily obtained by *substituting* $y = (1-x^2)$, and is $-\sqrt{(1-x^2)}$;

hence $\int_x \sqrt{\left(\dfrac{1+x}{1-x}\right)} = \sin^{-1} x - \sqrt{(1-x^2)}.$

(c) Required $\int_x \sqrt{(a^2 - x^2)}$.

$$\sqrt{(a^2-x^2)} = \frac{a^2-x^2}{\sqrt{(a^2-x^2)}} = \frac{a^2}{\sqrt{(a^2-x^2)}} - \frac{x^2}{\sqrt{(a^2-x^2)}},$$

and $\quad \int_x \sqrt{(a^2-x^2)} = a^2 \int_x \dfrac{1}{\sqrt{(a^2-x^2)}} - \int_x \dfrac{x^2}{\sqrt{(a^2-x^2)}}$,

whence $\quad \int_x \sqrt{(a^2-x^2)} = a^2 \sin^{-1}\dfrac{x}{a} - \int_x \dfrac{x^2}{\sqrt{(a^2-x^2)}}$, by Art. 77;

but $\quad \int_x \sqrt{(a^2-x^2)} = x\sqrt{(a^2-x^2)} + \int_x \dfrac{x^2}{\sqrt{(a^2-x^2)}}$,

by integration by parts, if we let

$$u = \sqrt{(a^2-x^2)} \text{ and } D_x v = 1.$$

Adding our two equations, we have

$$2\int_x \sqrt{(a^2-x^2)} = x\sqrt{(a^2-x^2)} + a^2 \sin^{-1}\dfrac{x}{a};$$

and $\quad \therefore \int_x \sqrt{(a^2-x^2)} = \dfrac{1}{2}\left(x\sqrt{a^2-x^2} + a^2 \sin^{-1}\dfrac{x}{a}\right).$

EXAMPLES.

Find

(1) $\int_x \sqrt{(x^2+a^2)}$.

$\quad$ Ans. $\dfrac{1}{2}[x\sqrt{(x^2+a^2)} + a^2 \log(x+\sqrt{x^2+a^2})]$.

(2) $\int_x \sqrt{(x^2-a^2)}$.

$\quad$ Ans. $\dfrac{1}{2}[x\sqrt{(x^2-a^2)} - a^2 \log(x+\sqrt{x^2-a^2})]$.

Applications.

83. *To find the area of a segment of a circle.*

Let the equation of the circle be

$$x^2 + y^2 = a^2,$$

and let the required segment be cut off by the double ordinates through (x_0, y_0) and (x, y). Then the required area

$$A = 2\int_x y + C.$$

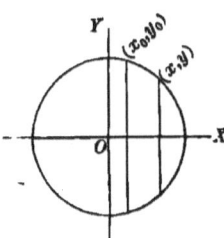

From the equation of the circle,
$$y = \sqrt{(a^2 - x^2)},$$
hence
$$A = 2\int_x \sqrt{(a^2 - x^2)} + C;$$
and therefore, by Art. 82 (c),
$$A = x\sqrt{(a^2 - x^2)} + a^2 \sin^{-1}\frac{x}{a} + C.$$

As the area is measured from the ordinate y_0 to the ordinate y,
$$A = 0 \text{ when } x = x_0;$$
therefore
$$0 = x_0\sqrt{(a^2 - x_0^2)} + a^2 \sin^{-1}\frac{x_0}{a} + C,$$
$$C = -x_0\sqrt{(a^2 - x_0^2)} - a^2 \sin^{-1}\frac{x_0}{a},$$
and we have
$$A = x\sqrt{(a^2 - x^2)} + a^2 \sin^{-1}\frac{x}{a} - x_0\sqrt{(a^2 - x_0^2)} - a^2 \sin^{-1}\frac{x_0}{a}$$

If $x_0 = 0$, and *the segment begins with the axis of* Y,
$$A = x\sqrt{(a^2 - x^2)} + a^2 \sin^{-1}\frac{x}{a}.$$

If, at the same time, $x = a$, *the segment becomes a semicircle*, and
$$A = a\sqrt{(a^2 - a^2)} + a^2 \sin^{-1}\frac{a}{a} = \frac{\pi a^2}{2}.$$

The area of the whole circle is πa^2.

Examples.

(1) Show that, in the case of an ellipse,

$$\frac{x^2}{a^2} + \frac{y^2}{b^2} = 1,$$

the area of a segment beginning with any ordinate y_0 is

$$A = \frac{b}{a}\left[x\sqrt{(a^2-x^2)} + a^2\sin^{-1}\frac{x}{a} - x_0\sqrt{(a^2-x_0^2)} - a^2\sin^{-1}\frac{x_0}{a} \right].$$

That if the segment begins with the minor axis,

$$A = \frac{b}{a}\left[x\sqrt{(a^2-x^2)} + a^2\sin^{-1}\frac{x}{a} \right].$$

That the area of the whole ellipse is πab.

(2) The area of a segment of the hyperbola

$$\frac{x^2}{a^2} - \frac{y^2}{b^2} = 1$$

is
$$A = \frac{b}{a}[x\sqrt{(x^2-a^2)} - a^2\log(x+\sqrt{x^2-a^2})$$
$$- x_0\sqrt{(x_0^2-a^2)} + a^2\log(x_0+\sqrt{x_0^2-a^2})].$$

If $x_0 = a$, and the segment begins at the vertex,

$$A = \frac{b}{a}[x\sqrt{(x^2-a^2)} - a^2\log(x+\sqrt{x^2-a^2}) + a^2\log a].$$

84. *To find the length of any arc of a circle, the coördinates of its extremities being (x_0, y_0) and (x, y).*

By Art. 52, $\quad s = \int_x \sqrt{[1 + (D_x y)^2]}.$

From the equation of the circle,

$$x^2 + y^2 = a^2,$$

CHAP. V.] INTEGRATION. 75

we have
$$2x + 2y D_x y = 0,$$

$$D_x y = -\frac{x}{y},$$

$$1 + (D_x y)^2 = \frac{x^2 + y^2}{y^2} = \frac{a^2}{y^2},$$

$$s = \int_x \frac{a}{y} = a \int_x \frac{1}{\sqrt{(a^2 - x^2)}} = a \sin^{-1}\frac{x}{a} + C. \quad \text{(Art. 77.)}$$

When $x = x_0,$ $s = 0;$

hence
$$0 = a \sin^{-1}\frac{x_0}{a} + C,$$

$$C = -a \sin^{-1}\frac{x_0}{a},$$

and
$$s = a\left(\sin^{-1}\frac{x}{a} - \sin^{-1}\frac{x_0}{a}\right).$$

If $x_0 = 0$, and *the arc is measured from the highest point of the circle,*
$$s = a \sin^{-1}\frac{x}{a}.$$

If the arc is a quadrant, $x = a,$

$$s = a \sin^{-1}(1) = \frac{\pi a}{2},$$

and the whole circumference $= 2\pi a.$

85. *To find the length of an arc of the parabola* $y^2 = 2mx.$

We have
$$2 y D_x y = 2m;$$

$$D_x y = \frac{m}{y};$$

$$\sqrt{[1 + (D_x y)^2]} = \sqrt{\left(\frac{m^2 + y^2}{y^2}\right)} = \frac{1}{y}\sqrt{(m^2 + y^2)};$$

$$s = \int_x \left[\frac{1}{y} \sqrt{(m^2 + y^2)} \right] = \int_y \left[\frac{1}{y} \sqrt{(m^2 + y^2)}\ D_y x \right];$$

$$D_y x = \frac{1}{D_x y} = \frac{y}{m}, \qquad \text{by Art. 73;}$$

$$s = \frac{1}{m} \int_y \sqrt{m^2 + y^2} = \frac{1}{2m} [y \sqrt{m^2 + y^2} + m^2 \log(y + \sqrt{m^2 + y^2})] + C,$$

by Art. 82, Ex. 1.

If the arc is measured from the vertex,

$$s = 0 \text{ when } y = 0;$$

$$0 = \frac{1}{2m}(m^2 \log m) + C,$$

$$C = -\frac{1}{2} m \log m,$$

and $\quad s = \dfrac{1}{2} \left[\dfrac{y\sqrt{(m^2 + y^2)}}{m} + m \log \dfrac{y + \sqrt{(m^2 + y^2)}}{m} \right].$

EXAMPLE.

Find the length of the arc of the curve $x^3 = 27 y^2$ included between the origin and the point whose abscissa is 15.

Ans. 19.

CHAPTER VI.

CURVATURE.

86. *The total curvature* of an arc of a continuous curve is its total change of direction, and is measured by the angle formed by the tangents at its extremities. The *mean curvature* of an arc is its total curvature divided by its length. The *actual curvature* of a curve at a given point is the limit approached by the mean curvature of the arc beginning at the point, as the length of the arc is indefinitely decreased.

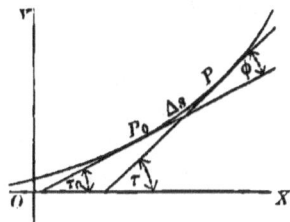

Thus, in the figure, the *total curvature* of the arc $P_0 P$, or Δs, is the angle φ, which is equal to $\tau - \tau_0$ or $\Delta \tau$. The *mean curvature* is $\dfrac{\Delta \tau}{\Delta s}$, and the *actual curvature* at P_0 is

$$\lim_{\Delta s \doteq 0} \left[\frac{\Delta \tau}{\Delta s} \right] = D_s \tau.$$

87. To find $D_s \tau$ in any particular example, we must, in theory, begin by expressing τ in terms of s by the aid of our old relations

$$\tan \tau = D_x y,$$

$$D_x s = \sqrt{[1 + (D_x y)^2]},$$

together with the equation of the given curve; but, in practice, this part of the work may be avoided. By the aid of the relations just referred to, τ and s may be expressed in terms of x; and, consequently, we may regard them as functions of x, and can obtain their derivatives with respect to x; and then the derivative of either with respect to the other may be found by the following principle.

88. If y is a function of x, and z is a function of x,

$$D_y z = \frac{D_x z}{D_x y}. \qquad [1]$$

For

$$\frac{D_x z}{D_x y} = \frac{\underset{\Delta x \doteq 0}{\text{limit}} \left[\frac{\Delta z}{\Delta x} \right]}{\underset{\Delta x \doteq 0}{\text{limit}} \left[\frac{\Delta y}{\Delta x} \right]} = \underset{\Delta x \doteq 0}{\text{limit}} \left[\frac{\frac{\Delta z}{\Delta x}}{\frac{\Delta y}{\Delta x}} \right]$$

$$= \underset{\Delta x \doteq 0}{\text{limit}} \left[\frac{\Delta z}{\Delta y} \right] = \underset{\Delta y \doteq 0}{\text{limit}} \left[\frac{\Delta z}{\Delta y} \right] = D_y z;$$

for Δx, Δy, and Δz approach zero simultaneously.

89. We have thus, if $\varkappa$ represents the curvature at any point of the curve,

$$\varkappa = D_s \tau = \frac{D_x \tau}{D_x s}.$$

Since

$$\tan \tau = D_x y,$$

$$\sec^2 \tau \, D_x \tau = D_x^2 y,$$

$$D_x \tau = \frac{D_x^2 y}{\sec^2 \tau};$$

but

$$\sec^2 \tau = 1 + \tan^2 \tau = 1 + (D_x y)^2,$$

and

$$D_x \tau = \frac{D_x^2 y}{1 + (D_x y)^2};$$

and, as
$$D_x s = \sqrt{[1+(D_x y)^2]},$$

we have
$$\varkappa = \frac{D_x^2 y}{\pm[1+(D_x y)^2]^{\frac{3}{2}}}.$$

Either the positive or the negative value might be chosen as the normal one. For reasons that will be evident hereafter, it is customary to use the negative one; and we have

$$\varkappa = \frac{-D_x^2 y}{[1+(D_x y)^2]^{\frac{3}{2}}}.$$

(a) For example, let it be required to find *the curvature of a straight line* $Ax+By+C=0$ at any point.

Differentiating with respect to x, we have

$$A+BD_x y = 0;$$

$$D_x y = -\frac{A}{B};$$

$$D_x^2 y = 0;$$

$$1+(D_x y)^2 = \frac{A^2+B^2}{B^2};$$

$$\varkappa = \frac{-D_x^2 y}{[1+(D_x y)^2]^{\frac{3}{2}}} = \frac{0}{\left(\frac{A^2+B^2}{B^2}\right)^{\frac{3}{2}}} = 0;$$

a result which might have been anticipated.

(b) *The curvature of a circle,*

$$x^2+y^2 = a^2.$$

$$2x+2yD_x y = 0;$$

$$D_x y = -\frac{x}{y};$$

$$D_x^2 y = -\frac{y - xD_x y}{y^2} = -\frac{y + \dfrac{x^2}{y}}{y^2} = -\frac{x^2 + y^2}{y^3} = -\frac{a^2}{y^3};$$

$$1 + (D_x y)^2 = \frac{x^2 + y^2}{y^2} = \frac{a^2}{y^2};$$

$$\varkappa = \frac{a^2}{y^3} \div \left(\frac{a^2}{y^2}\right)^{\frac{3}{2}} = \frac{a^2}{y^3} \div \frac{a^3}{y^3} = \frac{1}{a}.$$

Hence *the curvature of a circle is the same at every point, and is equal to the reciprocal of the radius.*

If
$$a = 1,$$
$$\varkappa = 1;$$

and the *unit of curvature is the curvature of the circle whose radius is unity.*

(c) The curvature of a parabola,

$$y^2 = 2mx.$$

$$2yD_x y = 2m;$$

$$D_x y = \frac{m}{y};$$

$$D_x^2 y = -\frac{m}{y^2} D_x y = -\frac{m^2}{y^3};$$

$$1 + (D_x y)^2 = \frac{m^2 + y^2}{y^2};$$

$$\varkappa = \frac{m^2}{y^3} \div \frac{(m^2 + y^2)^{\frac{3}{2}}}{y^3} = \frac{m^2}{(m^2 + y^2)^{\frac{3}{2}}};$$

and is a function of y, one of the coördinates of the point con-

sidered. From the form of $\varkappa$, it is obvious that the curvature is greatest when
$$y = 0;$$
that is, at the vertex of the curve; that it decreases as y increases or decreases, and that it has equal values for values of y which are equal with opposite signs.

EXAMPLES.

(1) Required the curvature of the ellipse
$$\frac{x^2}{a^2} + \frac{y^2}{b^2} = 1 \quad \text{at any point.}$$

Ans. $\varkappa = \dfrac{a^4 b^4}{(b^4 x^2 + a^4 y^2)^{\frac{3}{2}}}.$

(2) Of the hyperbola $\dfrac{x^2}{a^2} - \dfrac{y^2}{b^2} = 1.$

Ans. $\varkappa = \dfrac{a^4 b^4}{(b^4 x^2 + a^4 y^2)^{\frac{3}{2}}}.$

(3) Of the equilateral hyperbola
$$xy = \frac{a^2}{2}.$$

Ans. $\varkappa = -\dfrac{a^2}{(x^2 + y^2)^{\frac{3}{2}}}.$

Osculating Circle.

90. As the curvature of a circle has been found to be the reciprocal of its radius, a circle may be drawn which shall have any curvature required. *A circle tangent to a curve at any point, and having the same curvature as the curve at that point*, is called the *osculating circle* of the curve at the point in question. Its *radius* is called the *radius of curvature* of the curve at the point, and its *centre* is called the *centre of curvature*.

From the definition of the radius of curvature, it is obviously

normal to the curve, and its length is the reciprocal of the curvature at the point. If ρ represents the radius of curvature, we have
$$\rho = \frac{1}{x}.$$

Of course, ρ is generally a function of the coördinates of the point of the curve, and changes its length as the position of the point in question is changed.

Evolutes.

91. *The locus of the centre of curvature of a given curve is the evolute of the curve.*

PROBLEM.

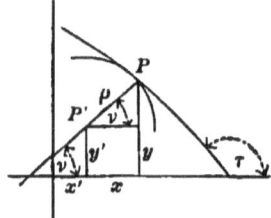

To find the equation of the evolute of a given curve
$$y = fx.$$

Let P, coördinates (x,y), be any point of the curve, and P', (x',y') the corresponding point of the evolute; ν the angle made by the normal with the axis of x, and ρ the radius of curvature at P. ρ and τ can be found from the equation of the curve, and
$$\nu = \tau - 90°.$$

We see from the figure, that
$$x' = x - \rho \cos \nu,$$
$$y' = y - \rho \sin \nu :$$

ρ and ν can be expressed in terms of x and y; and then, with the given equation,
$$y = fx,$$
we shall have three equations connecting the four variables, x, y, x', and y'. We can eliminate x and y, and so obtain a single equation connecting x' and y', the variable coördinates of any point on the evolute; and this will be the equation required.

92. For example: Let us find the *evolute of the parabola*
$$y^2 = 2mx.$$

$$\tan\tau = D_x y = \frac{m}{y};$$

$$\tan\nu = \tan(\tau - 90°) = -\cot\tau = -\frac{y}{m};$$

$$\sec^2\nu = 1 + \tan^2\nu = \frac{m^2 + y^2}{m^2};$$

$$\cos\nu = \pm \frac{m}{\sqrt{(m^2 + y^2)}};$$

$$\sin\nu = \pm \frac{y}{\sqrt{(m^2 + y^2)}}.$$

Since ν is given by its tangent, it may always be taken less than 180°; therefore we may take the positive value of $\sin\nu$, and in that case, as $\tan\nu$ is negative, we must take $\cos\nu$ with the negative sign: we have then

$$\sin\nu = \frac{y}{\sqrt{(m^2 + y^2)}};$$

$$\cos\nu = \frac{-m}{\sqrt{(m^2 + y^2)}}.$$

We have seen, Art. 89 (c), that

$$x = \frac{m^2}{(m^2 + y^2)^{\frac{1}{2}}};$$

hence
$$\rho = \frac{(m^2 + y^2)^{\frac{3}{2}}}{m^2};$$

$$x' = x + \frac{m^2 + y^2}{m};$$

$$y' = y - \frac{(m^2 + y^2) y}{m^2};$$

and these equations, together with

$$y^2 = 2mx,$$

are the equations of the evolute.

Reducing, we have $\quad x' = m + 3x;$

whence $\quad x = \dfrac{x' - m}{3};$

and $\quad y' = -\dfrac{y^3}{m^2};$

whence $\quad y = -(m^2 y')^{\frac{1}{3}}.$

Substituting in the equation of the parabola, we have

$$(m^2 y')^{\frac{2}{3}} = \frac{2m}{3} (x' - m).$$

Reducing, $\quad m^4 y'^2 = \dfrac{8}{27} m^3 (x' - m)^3,$

$$y'^2 = \frac{8}{27m} (x' - m)^3;$$

or, dropping accents,

$$y^2 = \frac{8}{27m} (x - m)^3,$$

the required evolute; a semi-cubical parabola.

93. By expressing ρ and ν in terms of x and y in the general equations of the evolute of $y = fx$, we can throw these equations into a rather more convenient form.

We have the values
$$\rho = -\frac{[1+(D_x y)^2]^{\frac{3}{2}}}{D_x^2 y},$$

$$\tan \tau = D_x y,$$

$$\tan \nu = -\frac{1}{D_x y},$$

and
$$\cot \nu = -D_x y.$$

$$\sin \nu = \frac{1}{[1+(D_x y)^2]^{\frac{1}{2}}},$$

$$\cos \nu = -\frac{D_x y}{[1+(D_x y)^2]^{\frac{1}{2}}},$$

$$x' = x - \frac{[1+(D_x y)^2]^{\frac{3}{2}}}{D_x^2 y} \cdot \frac{D_x y}{[1+(D_x y)^2]^{\frac{1}{2}}},$$

$$y' = y + \frac{[1+(D_x y)^2]^{\frac{3}{2}}}{D_x^2 y} \cdot \frac{1}{[1+(D_x y)^2]^{\frac{1}{2}}}.$$

Reducing
$$\left. \begin{aligned} x' &= x - D_x y \, \frac{1+(D_x y)^2}{D_x^2 y} \\ y' &= y + \frac{1+(D_x y)^2}{D_x^2 y} \end{aligned} \right\} \qquad [1]$$

EXAMPLE.

Required the evolute of a circle. *Ans.* $x' = 0$, $y' = 0$.

94. *To find the evolute of an ellipse,*

$$\frac{x^2}{a^2} + \frac{y^2}{b^2} = 1.$$

$$D_x y = \frac{-b^2 x}{a^2 y};$$

$$D_x^2 y = -\frac{b^4}{a^2 y^3};$$

$$1 + (D_x y)^2 = \frac{b^4 x^2 + a^4 y^2}{a^4 y^2};$$

$$x' = x + \frac{b^2 x}{a^2 y} \cdot \frac{b^4 x^2 + a^4 y^2}{a^4 y^2}\left(-\frac{a^2 y^3}{b^4}\right) = x - \frac{x(b^4 x^2 + a^4 y^2)}{a^4 b^2};$$

$$y' = y + \frac{b^4 x^2 + a^4 y^2}{a^4 y^2}\left(-\frac{a^2 y^3}{b^4}\right) = y - \frac{y(b^4 x^2 + a^4 y^2)}{a^2 b^4}.$$

Since
$$\frac{x^2}{a^2} + \frac{y^2}{b^2} = 1,$$

$$b^2 x^2 + a^2 y^2 = a^2 b^2;$$

$$a^4 y^2 = a^2 b^2(a^2 - x^2);$$

and
$$b^4 x^2 = a^2 b^2(b^2 - y^2).$$

$$b^4 x^2 + a^4 y^2 = b^2(a^4 - a^2 x^2 + b^2 x^2),$$

or
$$a^2(b^4 - b^2 y^2 + a^2 y^2).$$

$$x' = x - \frac{x(a^4 - a^2 x^2 + b^2 x^2)}{a^4} = \frac{a^2 - b^2}{a^4} x^3;$$

$$y' = y - \frac{y(b^4 - b^2 y^2 + a^2 y^2)}{b^4} = -\frac{a^2 - b^2}{b^4} y^3;$$

$$x = \left(\frac{a^4 x'}{a^2 - b^2}\right)^{\frac{1}{3}};$$

$$y = -\left(\frac{b^4 y'}{a^2 - b^2}\right)^{\frac{1}{3}}.$$

Substituting in
$$\frac{x^2}{a^2} + \frac{y^2}{b^2} = 1,$$

we have
$$\left(\frac{ax'}{a^2-b^2}\right)^{\frac{2}{3}} + \left(\frac{by'}{a^2-b^2}\right)^{\frac{2}{3}} = 1;$$

or, dropping accents,
$$(ax)^{\frac{2}{3}} + (by)^{\frac{2}{3}} = (a^2-b^2)^{\frac{2}{3}}.$$

EXAMPLE.

Find the evolute of the hyperbola $\dfrac{x^2}{a^2} - \dfrac{y^2}{b^2} = 1$.

Ans. $(ax)^{\frac{2}{3}} - (by)^{\frac{2}{3}} = (a^2+b^2)^{\frac{2}{3}}.$

Properties of the Evolute.

95. We have defined the evolute as the *locus of the centre of curvature* of the curve. It is also the *envelop of the normals* of the given curve, as may be readily shown; that is, *every normal to the curve is tangent to the evolute*. Let ν be the inclination of the normal at the point (x,y) of the given curve to the axis of X, and τ' the inclination of the tangent at the corresponding point (x',y') of the evolute. We have seen already that the normal at (x,y) passes through (x',y'), so it is only necessary to prove that $\tau' = \nu$.

But $\tan \tau' = D_{x'} y'$

and $\tan \nu = -\dfrac{1}{D_x y}.$

Hence we must show that $D_{x'} y' = -\dfrac{1}{D_x y}.$

By Art. 88, $D_{x'} y' = \dfrac{D_x y'}{D_x x'},$

since x' and y' may both be regarded as functions of x.

$$x' = x - \frac{D_x y [1 + (D_x y)^2]}{D_x^2 y};$$

$$y' = y + \frac{1 + (D_x y)^2}{D_x^2 y};$$

$$D_x x' = 1 - \frac{(D_x^2 y)^2 + 3(D_x y)^2 (D_x^2 y)^2 - D_x y [1 + (D_x y)^2] D_x^3 y}{(D_x^2 y)^2}$$

$$= -\frac{D_x y \{3 D_x y (D_x^2 y)^2 - [1 + (D_x y)^2] D_x^3 y\}}{(D_x^2 y)^2};$$

$$D_x y' = D_x y + \frac{2 D_x y (D_x^2 y)^2 - [1 + (D_x y)^2] D_x^3 y}{(D_x^2 y)^2}$$

$$= \frac{3 D_x y (D_x^2 y)^2 - [1 + (D_x y)^2] D_x^3 y}{(D_x^2 y)^2};$$

$$D_{x'} y' = \frac{D_x y'}{D_x x'} = -\frac{1}{D_x y}. \qquad \text{Q.E.D.}$$

96. A second important property of the evolute is that *the length of any arc of the evolute is the difference between the lengths of the radii of curvature of the given curve which pass through the extremities of the arc in question.*

Let (x_1', y_1') and (x_2', y_2') be the extremities of any arc of the evolute; ρ_1 and ρ_2 the radii of curvature drawn from these points to the curve; s_1' the length of the arc of the evolute measured from some fixed point on the evolute to (x_1', y_1'); and s_2' the length of an arc measured from the same fixed point to (x_2', y_2'). Then we wish to prove that

$$s_2' - s_1' = \rho_2 - \rho_1,$$

or
$$\Delta s' = \Delta \rho,$$

or
$$\frac{\Delta s'}{\Delta \rho} = 1,$$

or
$$\lim_{\Delta \rho \doteq 0} \left[\frac{\Delta s'}{\Delta \rho} \right] = 1,$$

or
$$D_\rho s' = 1,$$

where s' must be regarded as a function of ρ.

But
$$D_\rho s' = \frac{D_x s'}{D_x \rho}$$

and
$$D_x s' = \frac{D_{x'} s'}{D_{x'} x} = D_{x'} s' . D_x x'.$$

$$D_{x'} s' = [1 + (D_{x'} y')^2]^{\frac{1}{2}},$$

$$D_{x'} y' = -\frac{1}{D_x y}, \qquad \text{by Art. 95;}$$

hence
$$D_{x'} s' = \frac{1}{D_x y} [1 + (D_x y)^2]^{\frac{1}{2}}.$$

$$D_x s' = -\frac{[1+(D_x y)^2]^{\frac{1}{2}} \{3 D_x y (D_x^2 y)^2 - [1+(D_x y)^2] D_x^3 y\}}{(D_x^2 y)^2},$$
by Art. 95.

$$\rho = -\frac{[1+(D_x y)^2]^{\frac{3}{2}}}{D_x^2 y};$$

$$D_x \rho = \frac{-[1+(D_x y)^2]^{\frac{1}{2}} \{3 D_x y (D_x^2 y)^2 - [1+(D_x y)^2] D_x^3 y\}}{(D_x^2 y)^2};$$

$$D_\rho s' = \frac{D_x s'}{D_x \rho} = 1. \qquad \text{Q.E.D.}$$

97. These two properties enable us to *regard any curve as traced by the extremity of a stretched string unwound from the evolute*, the string being always tangent to the evolute, and its free portion at any instant being the radius of curvature of the curve at the point traced at that instant. *From this point of view, the curve itself* is called *the involute*.

CHAPTER VII.

SPECIAL EXAMPLES AND APPLICATIONS.

The Cycloid.

98. The *cycloid*, a plane curve possessing very remarkable geometrical and mechanical properties, was first studied just before the invention of the Calculus, and has always been a favorite with mathematicians.

It is the curve described in space by a fixed point in the rim of a wheel as the wheel rolls along in a straight line; or, more strictly, it is the curve described by any fixed point in the circumference of a circle, as the circle, keeping always in the same plane, rolls without sliding along a fixed straight line. The rolling circle is called the *generating circle*, and the fixed point the *generating point*, of the cycloid.

The curve will evidently consist of an indefinite number of equal arches, and can be cut by a straight line in an unlimited number of points. Its equation, then, cannot be of a finite degree, and so cannot be an algebraic equation. The curve is a *transcendental*, as distinguished from an *algebraic*, curve.

99. As the arches are all alike, it will do to consider a single one. Its base is obviously equal to the circumference, and its height to the diameter, of the generating circle, and its right and left hand halves are symmetrical.

We can get its equation most easily with the aid of an auxiliary angle. Take as axes the base of the cycloid, and a perpendicular to the base through the lowest position of the generating point, and represent by θ the angle made by the radius drawn to the generating point at any instant, with the radius drawn to the lowest point of the generating circle. The arc

SPECIAL EXAMPLES AND APPLICATIONS.

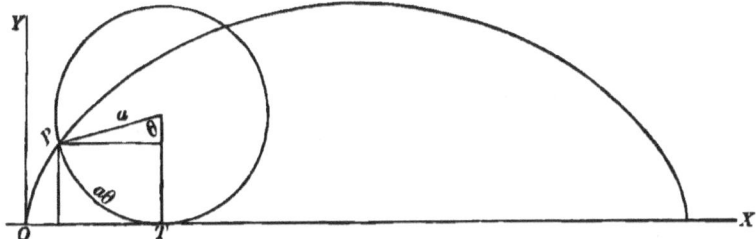

joining the two points just mentioned is $a\theta$, by Art. 66, if a is the radius of the circle; and this is therefore the length OT. If x and y are the coördinates of P, any point on the cycloid,

$$\left.\begin{array}{l} x = a\theta - a\sin\theta \\ y = a - a\cos\theta \end{array}\right\}; \qquad (A)$$

and these may be taken as the equations of the cycloid. Of course, θ may be eliminated between these equations, and a single equation obtained, containing x and y as the only variables. We get

$$\cos\theta = \frac{a-y}{a},$$

$$1 - \cos\theta = \frac{y}{a} = \text{vers}\,\theta,$$

hence

$$\theta = \text{vers}^{-1}\frac{y}{a};$$

$$\sin\theta = \sqrt{(1-\cos^2\theta)} = \pm\frac{1}{a}\sqrt{(2ay-y^2)},$$

and

$$x = a\,\text{vers}^{-1}\frac{y}{a} \mp \sqrt{(2ay-y^2)}, \qquad (B)$$

where the upper sign before the radical is to be used for points corresponding to values of $\theta < \pi$, that is, for points on the first half of the arch, and the lower sign for points on the second half of the curve.

Examples.

(1) Discuss completely the form of the cycloid from equations (A), supposing θ to increase from 0 to 2π.

(2) Discuss the form of the cycloid from equation (B), supposing y to increase from 0 to $2a$.

100. If our axes are lines parallel and perpendicular to the base through the highest point of the curve, the equations have a slightly different form. Let θ be measured from the highest point of the generating circle.

$$OT = AB = a\theta$$

and
$$\left.\begin{array}{l} x = a\theta + a\sin\theta \\ y = -a + a\cos\theta \end{array}\right\} \quad (C)$$

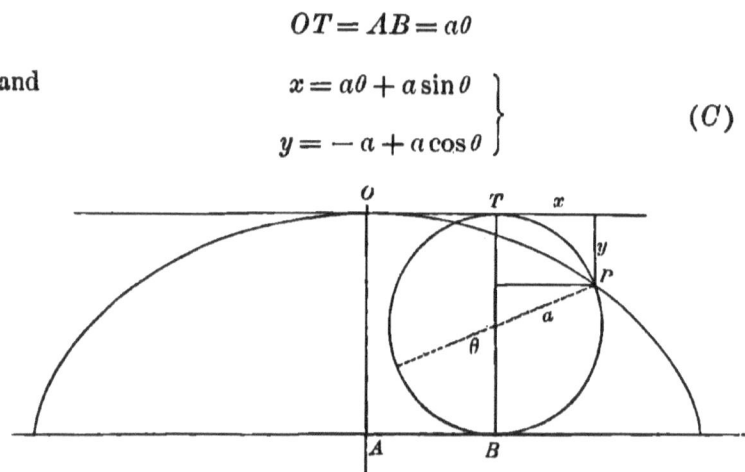

Examples.

(1) Obtain equations (C) from equations (A) by transformation of coördinates, noting that the formulas required are

$$x = a\pi + x',$$
$$y = 2a + y',$$
$$\theta = \pi + \theta'.$$

(2) Eliminate θ and obtain a single equation for the cycloid referred to its vertex as origin.

101. The properties of the curve can be investigated from the equations involving θ or from the single equation. In the text we shall employ the former. We ought to be able to determine (1) the direction of the tangent and normal at any point of the curve; (2) the equations of tangent and normal; (3) the lengths of tangent, normal, subtangent, and subnormal; (4) the curvature of the cycloid at any point; (5) the evolute; (6) the length of an arc of the curve; (7) the area of a segment of the curve.

(1) **102.**
$$x = a\theta - a\sin\theta,$$
$$y = a - a\cos\theta,$$
$$\tan\tau = D_x y = \frac{D_\theta y}{D_\theta x} = \frac{a\sin\theta}{a - a\cos\theta} = \frac{\sin\theta}{1 - \cos\theta} = \frac{2\sin\tfrac{\theta}{2}\cos\tfrac{\theta}{2}}{2\sin^2\tfrac{\theta}{2}} = \cot\tfrac{\theta}{2};$$
$$\tan\nu = -\cot\tau = -\tan\tfrac{\theta}{2}.$$

Since, as we have seen in Art. 99,
$$\sin\theta = \frac{1}{a}\sqrt{(2ay - y^2)}$$
and
$$1 - \cos\theta = \frac{y}{a},$$

$\tan\tau$ can be written $= \sqrt{\left(\dfrac{2a}{y} - 1\right)},$

and
$$\tan\nu = -\frac{y}{\sqrt{(2ay - y^2)}}.$$

Since
$$\tan\nu = -\tan\tfrac{\theta}{2},$$
$$\nu = \pi - \tfrac{\theta}{2} \qquad \text{by trigonometry.}$$

In the figure (see next page), PTO being formed by a tangent and a chord, is measured by half the arc PT, and therefore is equal to $\tfrac{\theta}{2}$. PTA, then, is equal to ν, and the line PT is a normal. Hence the *normal at any point on the cycloid passes through the lowest point of the generating circle*. The tangent must, therefore, pass through the highest point of the generating circle.

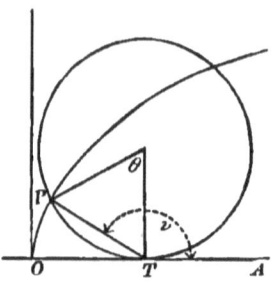

EXAMPLES.

(1) At what point of the curve is the tangent parallel to the base of the cycloid? Perpendicular to the base? Where does it make an angle of 45° with the base?

(2) Obtain the values of $\tan \tau$ and $\tan \nu$ from equation (B).

(2) 103. The equation of the tangent at the point (x_0, y_0) is

by Art. 28, $\qquad y - y_0 = \cot \frac{\theta_0}{2} (x - x_0),$

or $\qquad y - y_0 = \sqrt{\left(\dfrac{2a}{y_0} - 1\right)} (x - x_0) \ ;$

of the normal, is $\quad y - y_0 = - \tan \frac{\theta_0}{2} (x - x_0),$

or $\qquad y - y_0 = - \dfrac{y_0}{\sqrt{(2 a y_0 - y_0^2)}} (x - x_0).$

EXAMPLE.

Show, from the equation of the normal, that it passes through the point $(a\theta, 0)$, the lowest point of the generating circle.

(3) 104. We have the formulas,

$$t_x = \frac{y}{D_x y},$$

$$n_x = y D_x y,$$

$$t = \frac{y}{D_x y}\sqrt{[1+(D_x y)^2]},$$

$$n = y\sqrt{[1+(D_x y)^2]}; \qquad \text{(Art. 28)};$$

hence
$$t_x = \frac{a(1-\cos\theta)}{\cot\frac{\theta}{2}} = \frac{2a\sin^2\frac{\theta}{2}}{\cot\frac{\theta}{2}} = 2a\sin^2\frac{\theta}{2}\tan\frac{\theta}{2}.$$

$$n_x = a(1-\cos\theta)\cot\frac{\theta}{2} = 2a\sin^2\frac{\theta}{2}\cot\frac{\theta}{2} = 2a\sin\frac{\theta}{2}\cos\frac{\theta}{2} = a\sin\theta,$$

$$t = 2a\sin^2\frac{\theta}{2}\tan\frac{\theta}{2}\csc\frac{\theta}{2} = 2a\sin\frac{\theta}{2}\tan\frac{\theta}{2}.$$

$$n = 2a\sin^2\frac{\theta}{2}\csc\frac{\theta}{2} = 2a\sin\frac{\theta}{2}.$$

Since
$$D_x y = \sqrt{\left(\frac{2a}{y}-1\right)},$$

the value of n may be expressed,

$$n = \sqrt{(2ay)}.$$

(4) 105.
$$\varkappa = \frac{-D_x^2 y}{[1+(D_x y)^2]^{\frac{3}{2}}}; \qquad \text{(Art. 89)}.$$

$$D_x^2 y = \frac{D_\theta(D_x y)}{D_\theta x} = \frac{-\frac{1}{2}\csc^2\frac{\theta}{2}}{2a\sin^2\frac{\theta}{2}} = -\frac{1}{4a}\csc^4\frac{\theta}{2},$$

$$[1+(D_x y)^2]^{\frac{3}{2}} = \csc^3\frac{\theta}{2},$$

hence
$$\varkappa = \frac{1}{4a}\csc\frac{\theta}{2},$$

and
$$\rho = \frac{1}{\varkappa} = 4a\sin\frac{\theta}{2} = 2n = 2\sqrt{(2ay)};$$

and the *radius of curvature at any point is equal to twice the normal drawn at the point.*

EXAMPLES.

(1) Find at what points of the curve the curvature is greatest; at what least.

(2) Obtain the expression for the curvature from the equation (B).

(5) 106. The equations of the evolute of a curve are

$$x' = x - D_z y \frac{1+(D_z y)^2}{D_z^2 y}$$

$$y' = y + \frac{1+(D_z y)^2}{D_z^2 y}$$

(Art. 93 [1]).

Here

$$x' = a\theta - a\sin\theta - \frac{\cot\frac{\theta}{2}\csc^2\frac{\theta}{2}}{-\frac{1}{4a}\csc^4\frac{\theta}{2}} = a\theta - a\sin\theta + 4a\sin\frac{\theta}{2}\cos\frac{\theta}{2}$$

$$= a\theta - a\sin\theta + 2a\sin\theta,$$

$$= a\theta + a\sin\theta.$$

$$y' = a - a\cos\theta + \frac{\csc^2\frac{\theta}{2}}{-\frac{1}{4a}\csc^4\frac{\theta}{2}} = a - a\cos\theta - 4a\sin^2\frac{\theta}{2}$$

$$= a(1-\cos\theta) - 2a(1-\cos\theta) = -a + a\cos\theta;$$

and we have, as the equations of the evolute,

$$\left.\begin{array}{l} x' = a\theta + a\sin\theta \\ y' = -a + a\cos\theta \end{array}\right\};$$

but these (Art. 100) are the equations of an equal cycloid referred to the tangent and normal at the vertex as axes. The

CHAP. VII.] SPECIAL EXAMPLES AND APPLICATIONS. 97

cycloid and its evolute would be situated as indicated by the figure.

The property of the evolute established in Art. 96 enables us to obtain easily the length of the arc of an arch of the cycloid

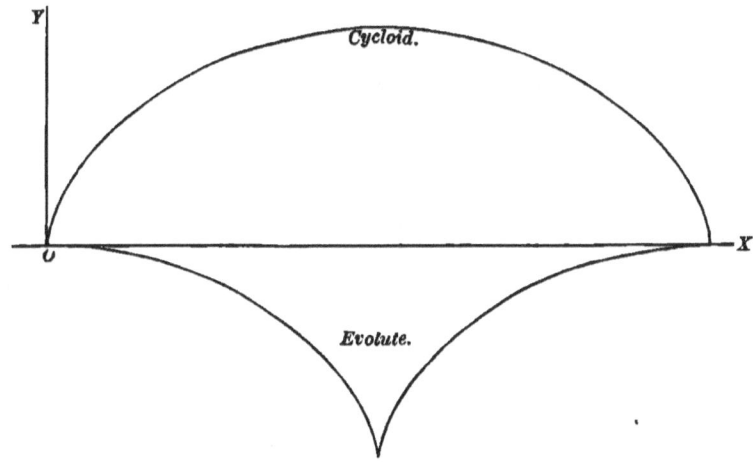

The length of the half-arch of the evolute is the difference between the radii of curvature at the highest and the lowest points of the given curve; that is,

$$[\rho]_{\theta=\pi} - [\rho]_{\theta=0} = 4a\sin\frac{\pi}{2} - 4a\sin 0 = 4a,$$

and S, the whole arc, $= 8a$.

(6) 107. The length of an arc of the cycloid can be found from the formula $\quad s = \int_x [1 + (D_x y)^2]^{\frac{1}{2}}$
without using the evolute.

We have $\qquad D_x y = \cot\frac{\theta}{2},$

$$[1 + (D_x y)^2]^{\frac{1}{2}} = \csc\frac{\theta}{2};$$

hence $\qquad s = \int_x \csc\frac{\theta}{2} = \int_\theta \csc\frac{\theta}{2} D_\theta x :$

but $\qquad D_\theta x = 2a\sin^2\frac{\theta}{2},$

and $\qquad s = 2a\int_\theta \sin\frac{\theta}{2}.$

Let
$$z = \tfrac{\theta}{2};$$
then
$$D_z\theta = 2,$$
$$s = 2a\!\int_\theta \sin z = 2a\!\int_z \sin z\, D_z\theta = -4a\cos z + C,$$
$$s = -4a\cos\tfrac{\theta}{2} + C.$$

If we measure the arc from the origin, s must equal 0 when
$$\theta = 0.$$
$$0 = -4a\cos 0 + C,$$
$$C = 4a,$$
and we have $\quad s = 4a(1 - \cos\tfrac{\theta}{2}).$

To get the whole arch, let $\theta = 2\pi$,
$$s = 4a(1 - \cos\pi) = 8a.$$

(7) 108. For the area of a segment of the arch, we have the formula
$$A = \int_x y + C.$$
$$\int_x y = a\!\int_x(1 - \cos\theta) = a\!\int_\theta(1 - \cos\theta)D_\theta x = a^2\!\int_\theta(1 - \cos\theta)^2$$
$$= a^2\!\int_\theta(1 - 2\cos\theta + \cos^2\theta) = a^2(\int_\theta 1 - 2\!\int_\theta \cos\theta + \int_\theta \cos^2\theta),$$
$$\int_\theta 1 = \theta,$$
$$\int_\theta \cos\theta = \sin\theta,$$
$$\int_\theta \cos^2\theta = \tfrac{1}{2}(\theta + \sin\theta\cos\theta)$$
[see Art. 80, Ex. (1)];

hence $\quad A = a^2[\theta - 2\sin\theta + \tfrac{1}{2}(\theta + \sin\theta\cos\theta)] + C.$

If the segment is measured from the origin,
$$A = 0 \text{ when } \theta = 0;$$
$$0 = a^2[0 - 0 + \tfrac{1}{2}(0 + 0)] + C$$
and $\quad C = 0.$

CHAP. VII.] SPECIAL EXAMPLES AND APPLICATIONS. 99

The area of the whole arch is obtained by making

$$\theta = 2\pi.$$

$$A = a^2 [2\pi - 2\sin 2\pi + \tfrac{1}{2}(2\pi + \sin 2\pi \cos 2\pi)] = 3\pi a^2,$$

so that the area of the arch is three times the area of the generating circle.

EXAMPLE.

Find the length of an arc and the area of a segment from the equation (B).

109. *If the generating circle rolls on the circumference of a fixed circle,* instead of on a fixed line, the curve generated is called an *epicycloid,* if the rolling circle and the fixed circle are tangent *externally,* a *hypocycloid,* if they are tangent *internally.* The equations of these curves may be readily obtained. Let the

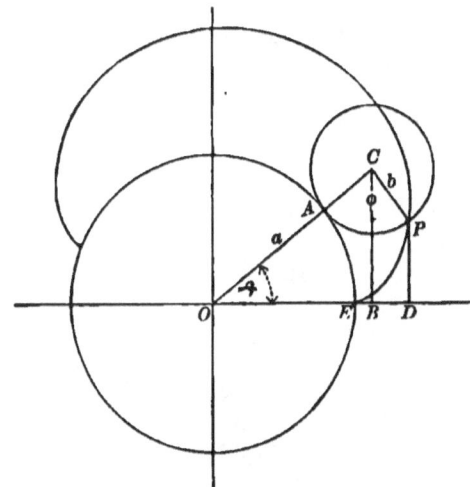

figure represent the generation of an epicycloid, P being the generating point and E the starting point. Call AOB, θ; and PCA, φ; OD is x and DP is y. Let a and b be the radii of fixed and rolling circles. Then

$$x = (a+b)\cos\theta + b\sin\left[\varphi - \left(\frac{\pi}{2} - \theta\right)\right],$$

$$y = (a+b)\sin\theta - b\cos\left[\varphi - \left(\frac{\pi}{2} - \theta\right)\right];$$

but the arcs AP and AE are equal, and

$$AP = b\varphi,$$
$$AE = a\theta,$$

hence $\quad a\theta = b\varphi$

and $\quad \varphi = \dfrac{a}{b}\theta;$

$$\varphi + \theta = \frac{a+b}{b}\theta;$$

and the equations become

$$\left.\begin{aligned} x &= (a+b)\cos\theta - b\cos\frac{a+b}{b}\theta \\ y &= (a+b)\sin\theta - b\sin\frac{a+b}{b}\theta \end{aligned}\right\}. \quad [1]$$

The equations of the hypocycloid are, in like manner, found to be

$$\left.\begin{aligned} x &= (a-b)\cos\theta + b\cos\frac{a-b}{b}\theta \\ y &= (a-b)\sin\theta - b\sin\frac{a-b}{b}\theta \end{aligned}\right\}. \quad [2]$$

EXAMPLES.

(1) If $b = a$ in the epicycloid, the curve is called a cardioide. Show that its polar equation is

$$r = 2a\,(1 - \cos\varphi)$$

when the starting point is taken as pole.

(2) If $a = 4b$ in the hypocycloid, obtain the cartesian equation of the curve by eliminating θ. *Ans.* $x^{\frac{2}{3}} + y^{\frac{2}{3}} = a^{\frac{2}{3}}$.

(3) If $a = 2b$ in the hypocyloid, show that the curve reduces to a diameter of the fixed circle.

(4) Prove by differentiation that the normal at any point of either epicycloid or hypocycloid passes through the point of contact of fixed and generating circles.

CHAPTER VIII.

PROBLEMS IN MECHANICS.

110. We have seen (Art. 12) that, if s represents the distance traversed by a moving body in t seconds, and can be expressed as a function of t, the velocity of the body at any instant
$$v = D_t s.$$

111. The *acceleration* of a moving body at any instant is the rate at which its velocity is changing at that instant. If the velocity is increasing, the acceleration is positive; if diminishing, the acceleration is negative. We shall represent it by a, and it is evidently a function of t. Since the derivative of a function measures the rate at which its value is changing (Art. 38), we shall have
$$a = D_t v = D_t^2 s,$$
since
$$v = D_t s.$$

For example: in the case of a body falling freely near the surface of the earth, we have approximately the law
$$s = 16 t^2.$$
Here
$$v = D_t s = 32 t,$$
and
$$a = D_t v = D_t^2 s = 32,$$
and the acceleration is constant and is equal to 32 feet a second; that is, the velocity of the fall at any instant is 32 feet a second greater than it was a second before. The relations
$$v = D_t s$$
and
$$a = D_t v = D_t^2 s,$$

and the corresponding formulas,

$$v = f_t a + C,$$

$$s = f_t v + C,$$

obtained by integrating them, are of great importance in problems concerning motion.

112. We shall assume the following principles of mechanics: (1) *A force acting on a body in the line of its motion produces an acceleration proportional to the intensity of the force;* and this acceleration is taken as the measure of the force. We speak of a force as *a force producing an acceleration of so many feet a second;* or, more briefly, as *a force of so many feet a second.* (2) *The effect of a force in producing acceleration in any direction not its own, is the product of the magnitude of the force by the cosine of the angle between the two directions;* or, in other words, it is the projection of the line representing the force in direction and intensity upon the line of the direction in question.

Problem.

113. The force exerted by the earth's attraction upon any particle of matter is constant at any given part of the earth's surface, and is nearly equal to 32 feet a second. Let g represent the exact value of this force at any given point of the earth's surface, required the velocity of a falling body at the end of t seconds, and the distance fallen in t seconds. Here a is constant and equal to g.

$$v = f_t a = f_t g = gt + C.$$

If the body falls from rest, its velocity is 0 when t is 0;

$$0 = g \times 0 + C,$$

$$C = 0;$$

and

$$v = gt.$$

$$s = f_t v = f_t gt = g f_t t = \tfrac{1}{2} gt^2 + C.$$

When t is 0, the distance fallen must be 0;

$$0 = \tfrac{1}{2} g \times 0 + C,$$

$$C = 0,$$

and
$$s = \tfrac{1}{2} gt^2.$$

If the body, instead of being dropped, had started with an initial velocity v_0, — for example, if it had been fired from a gun directly down or directly up, — we should have found a different value for C in the expression for the velocity,

$$v = gt + C;$$

for now, when
$$t = 0,$$

$$v = v_0;$$

hence
$$v_0 = g \times 0 + C,$$

$$C = v_0,$$

and
$$v = gt + v_0.$$

$$s = f_t v = f_t(gt + v_0) = \tfrac{1}{2} gt^2 + v_0 t + C;$$

but as
$$s = 0 \text{ when } t = 0,$$

$$C = 0,$$

and
$$s = \tfrac{1}{2} gt^2 + v_0 t.$$

114. The equation
$$a = g$$

or
$$D_t^2 s = g$$

can be integrated by a second method of considerable interest and generality. Multiply both members by $2 D_t s$.

$$2 D_t s \, D_t^2 s = 2g \, D_t s;$$

but
$$2 D_t s D_t^2 s = D_t (D_t s)^2;$$

hence
$$\int_t 2 D_t s D_t^2 s = (D_t s)^2,$$

and we have
$$(D_t s)^2 = 2g \int_t D_t s = 2gs + C$$

or
$$v^2 = 2gs + C.$$

In the case of a falling body, when
$$t = 0, \; v = 0, \text{ and } s = 0;$$

hence
$$C = 0$$

and
$$v^2 = 2gs,$$
$$v = \sqrt{(2gs)}, \qquad [1]$$

or
$$D_t s = \sqrt{(2gs)}.$$

We cannot integrate directly here, for the first member is a function of t and the second member a function of s; but since

$$D_t s = \frac{1}{D_s t}, \qquad \text{by Art. 73,}$$

$$D_s t = \frac{1}{\sqrt{(2gs)}} = \frac{1}{\sqrt{(2g)}} s^{-\frac{1}{2}}.$$

$$t = \int_s \frac{1}{\sqrt{(2g)}} s^{-\frac{1}{2}} = \frac{2}{\sqrt{(2g)}} s^{\frac{1}{2}} + C = \sqrt{\left(\frac{2s}{g}\right)} + C.$$

Since $s = 0$ when $t = 0$,
$$C = 0$$

and
$$t = \sqrt{\left(\frac{2s}{g}\right)}. \qquad [2]$$

It is easily seen that these new values for v and t are entirely consistent with those obtained in the last article.

115. If the force is any constant force f, instead of g, we have merely to substitute f for g in the preceding results. For example, take the case of a body sliding without friction down an inclined plane. Here, by Art. 112, (2), the accelerating force in the direction of the motion is $g\cos(90° - \varphi)$, therefore

$$a = g\sin\varphi,$$

$$v = \sqrt{(2g\sin\varphi \cdot s)},$$

and

$$t = \sqrt{\left(\frac{2s}{g\sin\varphi}\right)}$$

when there is no initial velocity. In this case, the velocity and time are easily expressed in terms of the vertical distance through which the body has descended. Let OP be s, and OA, the vertical distance, be y. Then

$$y = s\sin\varphi,$$

$$v = \sqrt{(2gy)},$$

and

$$t = \sqrt{\left(\frac{2s^2}{gs\sin\varphi}\right)} = s\sqrt{\left(\frac{2}{gy}\right)}.$$

Substitute y for s in Art. 114, [1] and [2], and we get, as the velocity the body would acquire falling freely through the vertical distance y, and the time required for the fall,

$$v = \sqrt{(2gy)},$$

$$t = \sqrt{\left(\frac{2y}{g}\right)}.$$

We see the two velocities are identical; that is, *the velocity acquired by a body descending an inclined plane is precisely what it would have acquired falling through the vertical distance it has actually descended.*

$\frac{s}{t}$ is the mean velocity of the body during its descent, and

$$\frac{s}{t} = \sqrt{\left(\frac{gy}{2}\right)} \quad \text{for the inclined plane,}$$

$$\frac{y}{t} = \sqrt{\left(\frac{gy}{2}\right)} \quad \text{for the falling body.}$$

Hence the *mean velocity of a body descending an inclined plane is equal to the mean velocity of a body which has fallen freely the same vertical distance.*

116. Let the figure represent a vertical circle. The time of descent of a body sliding down any chord is

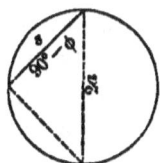

$$t = \sqrt{\left(\frac{2s}{g \sin \varphi}\right)} = \sqrt{\left(\frac{2s \csc \varphi}{g}\right)} = \sqrt{\left[\frac{2s \sec(90° - \varphi)}{g}\right]}$$

by Art. 115. If a is the radius,

$$s \cdot \sec(90° - \varphi) = 2a$$

and
$$t = 2\sqrt{\left(\frac{a}{g}\right)},$$

which is also the time a body would require to fall vertically the distance $2a$. Therefore, the time of descent down a chord of a vertical circle from the highest point of the circle to any point of the circumference is constant, and is equal to the time it would take the body to fall from the highest to the lowest point of the same circle.

Example.

Show that the time of descent down a chord from any point of a vertical circle to the lowest point of the circle is constant.

Problem.

117. To find the velocity acquired by a body falling from a distance toward the earth under the influence of the earth's attraction.

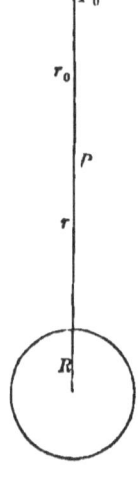

Here we cannot regard the attracting force as constant, as we do in dealing with small distances near the surface of the earth, but must take it as inversely proportional to the square of the distance of the body from the centre of the earth. Let R be the radius of the earth; r_0 the distance from the centre of the earth to the point at which the body started; r the distance from the centre to the position of the falling body when the time t has elapsed. Let g be the force of the attraction of the earth at the earth's surface, and f the force exerted at P.

Then we have
$$\frac{f}{g} = \frac{R^2}{r^2}$$

or
$$f = \frac{gR^2}{r^2} = a.$$

s, the distance fallen in the time t, equals $r_0 - r$.

$$D_t s = - D_t r = v,$$

$$D_t^2 s = - D_t^2 r = a;$$

hence
$$- D_t^2 r = \frac{gR^2}{r^2},$$

$$D_t^2 r = - \frac{gR^2}{r^2}.$$

Multiply by $2 D_t r$; $2 D_t r \, D_t^2 r = \dfrac{-2gR^2 D_t r}{r^2}$.

Integrate:
$$(D_t r)^2 = -2gR^2 \int_t \dfrac{D_t r}{r^2} = -2gR^2 \int_r \dfrac{1}{r^2} = \dfrac{2gR^2}{r} + C;$$

and
$$v^2 = \dfrac{2gR^2}{r} + C.$$

When the body was on the point of starting, its velocity was zero; hence, when $r = r_0$, $v = 0$;

and
$$0 = \dfrac{2gR^2}{r_0} + C,$$

$$C = -\dfrac{2gR^2}{r_0},$$

and
$$v^2 = 2gR^2 \left(\dfrac{1}{r} - \dfrac{1}{r_0} \right).$$

When the body reaches the surface of the earth,
$$r = R$$

and
$$v^2 = 2gR^2 \left(\dfrac{1}{R} - \dfrac{1}{r_0} \right).$$

The greater the value of r_0 in this result — that is, the greater the distance of the starting point from the centre of the earth — the nearer $\dfrac{1}{r_0}$ comes to the value 0, and the nearer v^2 approaches to $\dfrac{2gR^2}{R}$ or to $2gR$. In other words, the limiting value of the velocity acquired by a body falling from a distance to the surface of the earth under the influence of the earth's attraction, as the distance of the starting point is indefinitely increased, is $\sqrt{(2gR)}$.

Let us compute roughly the numerical value of this expression. g is about 32 feet per second; and as we use the foot as a unit in one of our values, we must in all: therefore R must be expressed in feet. R is about 4,000 miles, or 21,120,000 feet. $\sqrt{(R)} = 4,600$, nearly.

$$\sqrt{(2g)} = \sqrt{(64)} = 8.$$

$\sqrt{(2gR)} = 36,800$ feet, or nearly seven miles; and our required velocity is nearly seven miles a second; and neglecting the resistance of the air, this is the velocity with which a projectile would have to be thrown from the surface of the earth to prevent its returning.

We can easily go on and get an expression for the time of the fall by a second integration.

We have $(D_t r)^2 = 2gR^2 \left(\dfrac{1}{r} - \dfrac{1}{r_0} \right) = 2gR^2 \dfrac{r_0 - r}{r_0 r}$;

$$- D_t r = \sqrt{\left(\dfrac{2gR^2}{r_0} \right)} \sqrt{\left(\dfrac{r_0 - r}{r} \right)};$$

$$- D_r t = \sqrt{\left(\dfrac{r_0}{2gR^2} \right)} \sqrt{\left(\dfrac{r}{r_0 - r} \right)} \qquad \text{by Art. 73,}$$

$$- t = \sqrt{\left(\dfrac{r_0}{2gR^2} \right)} \int_r \sqrt{\left(\dfrac{r}{r_0 - r} \right)} + C,$$

$$\sqrt{\left(\dfrac{r}{r_0 - r} \right)} = \dfrac{r}{\sqrt{(r_0 r - r^2)}},$$

an expression to which we can apply the method of *integration by parts*.

Let $\qquad u = r,$

then $\qquad D_r u = 1;$

and let $\qquad D_r v = \dfrac{1}{\sqrt{(r_0 r - r^2)}},$

then $\qquad v = \text{vers}^{-1} \dfrac{2r}{r_0} \qquad$ by Art. 77, Ex. (2).

$$\int_r \dfrac{r}{\sqrt{(r_0 r - r^2)}} = r \, \text{vers}^{-1} \dfrac{2r}{r_0} - \int_r \text{vers}^{-1} \dfrac{2r}{r_0}$$

by Art. 79, [1].

Let $z = \dfrac{2r}{r_0}$,

then $D_z r = \dfrac{r_0}{2}$.

$$\int_r \operatorname{vers}^{-1} \dfrac{2r}{r_0} = \dfrac{r_0}{2} \int_z \operatorname{vers}^{-1} z = \dfrac{r_0}{2}\left[(z-1)\operatorname{vers}^{-1} z + \sqrt{(2z - z^2)}\right]$$

by Art. 75, [1], and Art. 81, Ex. (3). Replacing z by its value, $\int_r \operatorname{vers}^{-1} \dfrac{2r}{r_0} = \left(r - \dfrac{r_0}{2}\right)\operatorname{vers}^{-1}\dfrac{2r}{r_0} + \sqrt{(r_0 r - r^2)}.$

Whence $-t = \sqrt{\left(\dfrac{r_0}{2gR^2}\right)}\left[\dfrac{r_0}{2}\operatorname{vers}^{-1}\dfrac{2r}{r_0} - \sqrt{(r_0 r - r^2)}\right] + C.$

When $r = r_0,\ t = 0$;

hence $0 = \sqrt{\left(\dfrac{r_0}{2gR^2}\right)}\left(\dfrac{\pi r_0}{2}\right) + C,$

$$C = -\sqrt{\left(\dfrac{r_0}{2gR^2}\right)}\left(\dfrac{\pi r_0}{2}\right),$$

and $-t = \sqrt{\left(\dfrac{r_0}{8gR^2}\right)}\left[r_0\left(\operatorname{vers}^{-1}\dfrac{2r}{r_0} - \pi\right) - 2\sqrt{(r_0 r - r^2)}\right].$

EXAMPLES.

(1) The mean distance of the moon from the earth being 237,000 miles, find the velocity a body would acquire, and the time it would occupy, in falling from the moon to the earth's surface, neglecting the retarding effect of the moon's attraction.

(2) The force of the sun's attraction at its own surface is 905.5 feet; find the velocity a body would acquire, and the time it would occupy, in falling from the earth to the sun. Earth's mean distance = 92,000,000 miles; sun's diameter = 860,000 miles.

(3) Find the limit of the velocity a body could acquire falling from a distance to the sun.

(4) How long would it take Saturn to fall to the sun, Saturn's mean distance being about 880,000,000 miles?

PROBLEM.

118. To find the velocity acquired under the influence of gravity by a body sliding without friction down a given curve, or in any way constrained to move in a fixed curve.

Here the *effective* accelerating force is always tangent to the curve at the point the moving particle has reached. Suppose

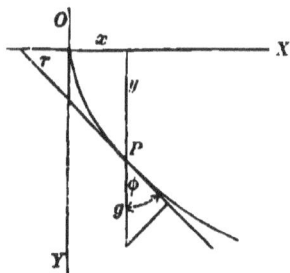

the origin of coördinates at the starting point, and let the direction downward be the positive direction of the ordinates. Of course, this will amount to changing the sign of $D_x y$; that is, will make τ the supplement of its usual value. The acceleration

$$a = g \cos \varphi = g \cos(90° - \tau) = g \sin \tau.$$

$$D_x y = \tan \tau,$$

$$1 + (D_x y)^2 = \sec^2 \tau,$$

$$\frac{1}{1 + (D_x y)^2} = \cos^2 \tau,$$

$$\frac{D_x y}{[1 + (D_x y)^2]^{\frac{1}{2}}} = \sin \tau,$$

hence
$$a = D_t^2 s = \frac{gD_x y}{[1+(D_x y)^2]^{\frac{1}{2}}};$$

but
$$[1+(D_x y)^2]^{\frac{1}{2}} = D_x s;$$

$$D_t^2 s = g\frac{D_x y}{D_x s} = gD_t y.$$

Multiply by $2 D_t s$:

$$2 D_t s \, D_t^2 s = 2gD_t y D_t s = 2gD_t y.$$

Integrate with respect to t, and

$$v^2 = (D_t s)^2 = 2gy + C.$$

If the particle started from rest at O,

$$v = 0 \text{ when } y = 0,$$

and
$$C = 0,$$

$$v = \sqrt{(2gy)};$$

but this is precisely the velocity it would have acquired in falling freely through the vertical distance y (Art. 114, [1]). So we are led to the remarkable result, that the velocity of a material particle, sliding without friction down a curve, under the influence of gravity, is the same at any instant as if it had fallen freely to the same vertical distance below the starting point. A special case of this has already been noticed in Art. 115.

Example.

Prove, from the equation of a circle, and the equation of a chord through its highest point, that the time of descent is independent of the length of the chord.

Problem.

119. To find the time of descent of a particle from any point of the arc of an inverted cycloid to the vertex of the curve. Taking the origin at the vertex of the curve, its equations are

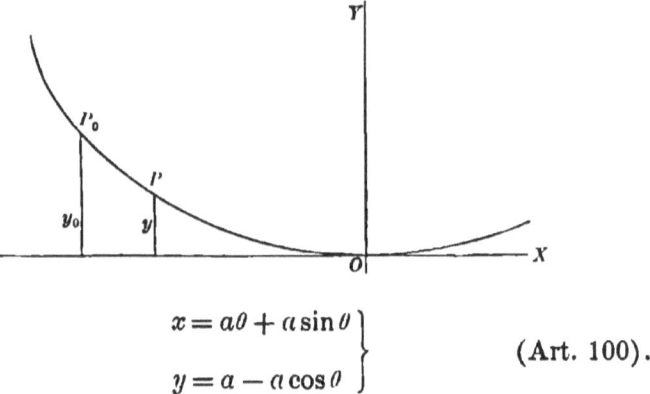

$$x = a\theta + a\sin\theta \atop y = a - a\cos\theta \Bigg\} \quad \text{(Art. 100).}$$

Let y_0 be the ordinate of the starting point, and y the ordinate of the point reached after t seconds. Then the vertical distance fallen is $y_0 - y$, and $\quad v = \sqrt{2g(y_0 - y)}, \quad$ by Art. 118.

$$D_t(s_0 - s) = - D_t s = \sqrt{2g(y_0 - y)} \,;$$

$$- D_s t = \frac{1}{\sqrt{2g(y_0 - y)}} \,;$$

$$- t = \int_s \frac{1}{\sqrt{2g(y_0 - y)}} = \int_y \frac{1}{\sqrt{2g(y_0 - y)}} \, D_y s \,;$$

$$D_x s = \sqrt{1 + (D_x y)^2} \,;$$

$$D_y s = \frac{D_x s}{D_x y} = \sqrt{\left(\frac{1}{D_x y}\right)^2 + 1} = \sqrt{1 + (D_y x)^2} \,;$$

$$D_\theta x = a + a\cos\theta \,;$$

$$D_\theta y = a\sin\theta \,;$$

$$D_y x = \frac{D_\theta x}{D_\theta y} = \frac{1 + \cos\theta}{\sin\theta} \,;$$

$$\cos\theta = \frac{a - y}{a} \,;$$

$$1 + \cos\theta = \frac{2a-y}{a};$$

$$\sin^2\theta = 1 - \cos^2\theta = \frac{a^2 - a^2 + 2ay - y^2}{a^2} = \frac{2ay - y^2}{a^2};$$

$$\sin\theta = \frac{1}{a}\sqrt{(2ay - y^2)};$$

$$D_y x = \frac{2a-y}{\sqrt{(2ay-y^2)}} = \sqrt{\frac{(2a-y)^2}{(2a-y)y}} = \sqrt{\left(\frac{2a-y}{y}\right)};$$

$$1 + (D_y x)^2 = \frac{2a}{y};$$

$$D_y s = \sqrt{\left(\frac{2a}{y}\right)};$$

$$-t = \int_y \frac{\sqrt{\left(\frac{2a}{y}\right)}}{\sqrt{2g(y_0 - y)}} = \sqrt{\left(\frac{a}{g}\right)} \int_y \frac{1}{\sqrt{(y_0 y - y^2)}} = \sqrt{\left(\frac{a}{g}\right)} \operatorname{vers}^{-1} \frac{2y}{y_0} + C,$$

by Art. 77 (2). When $y = y_0$, $t = 0$;

hence
$$0 = \sqrt{\left(\frac{a}{g}\right)} \operatorname{vers}^{-1}(2) + C.$$

$\operatorname{vers}^{-1}(2)$ is the angle which has the cosine -1, that is, the angle π. Hence,
$$C = -\sqrt{\left(\frac{a}{g}\right)} \pi,$$

and
$$-t = \sqrt{\left(\frac{a}{g}\right)} \left(\operatorname{vers}^{-1} \frac{2y}{y_0} - \pi\right).$$

When the particle reaches the vertex,

$$y = 0,$$

$$\operatorname{vers}^{-1} \frac{2y}{y_0} = 0,$$

and
$$t = \pi \sqrt{\left(\frac{a}{g}\right)}.$$

As this expression is independent of y_0, *the ordinate of the starting point, the time of descent to the vertex will be the same for all points of the curve.* If a pendulum were made to swing in a cycloid, this time $\pi\sqrt{\left(\dfrac{a}{g}\right)}$ would be one-half the time of a complete vibration, which would therefore be independent of the length of the arc. On account of this property, the cycloid is called the *tautochrone curve*.

EXAMPLE.

120. It is shown in mechanics, that, if the earth were a perfect and homogeneous sphere, and a cylindrical hole having its axis coincident with a diameter were bored through it, the attraction exerted on any body within this opening would be proportional to its distance from the centre. Find the expression for the velocity of a body at any instant, supposing it to have been dropped into this hole, and the time it would take to reach any given point of its course. Compute (1) its velocity when half-way to the centre; (2) when at the centre; (3) the time it would take it to reach the centre, if dropped from the surface; (4) if dropped from any point below the surface. Given $g = 32$; R, the radius of the earth, $= 4,000$ miles.

CHAPTER IX.

DEVELOPMENT IN SERIES.

121. A *series* is a sum composed of an unlimited number of terms which follow one another according to some law. If the terms of a series are real and finite, the sum of the first n terms is a definite value, no matter how great the value of n. If this sum *approaches a definite limit as* n *is indefinitely increased*, the series is *convergent; if not*, it is *divergent*. The limit approached by the sum of the first n terms of a convergent series as n increases indefinitely, is called the *sum of the series*, or simply the *series*. Thus, we may express the result arrived at in Art. 6 by saying the *sum of the series* $1 + \frac{1}{2} + \frac{1}{4} + \frac{1}{8} + \cdots$ is 2; or, more briefly, the *series* $1 + \frac{1}{2} + \frac{1}{4} + \frac{1}{8} + \cdots = 2$.

EXAMPLE.

122. Take the series $1 + x + x^2 + x^3 + \cdots$, *ad infinitum*. The series is a *geometrical progression*, and the sum of n terms can be found by the formula $s = \dfrac{ar^n - a}{r - 1}$.

Here $s = \dfrac{x^n - 1}{x - 1} = \dfrac{1 - x^n}{1 - x}$.

If $x < 1$, $\displaystyle\lim_{n = \infty} [x^n] = 0$,

and the sum of the series $= \dfrac{1}{1 - x}$, a definite value, and the series is, therefore, convergent.

If $x>1$, $\qquad x^n = \infty$ when $n = \infty$,

and the sum increases without limit as the number of terms increases indefinitely, and the series is divergent. The series $1 + x + x^2 + x^3 + \cdots$ can be obtained from $\dfrac{1}{1-x}$ by actual division, *but the fraction and the series are equal only when* $x<1$; for $\dfrac{1}{1-x}$ has a definite value when $x>1$; but, as we have seen, the series in that case has not a definite sum. *It is very unsafe to make use of divergent series, or to base any reasoning upon them,* for, from their nature, they are wholly indefinite. Convergent series, on the other hand, are perfectly definite values.

It is easily seen that the sum of the first n terms of a series *cannot approach indefinitely a fixed value* as n is increased, *unless, as we advance in the series, the terms eventually decrease;* or, in other words, *unless the ratio of the nth term to the one before it eventually becomes and remains less than unity* as n is increased. This, however, affords only a negative test for the convergency of series, as a series may not be convergent even when each term is less than the term before it.

123. The series we have just considered is an example of a series arranged according to the ascending powers of a variable, and such series play an important part in the theory of functions. We are naturally led to the consideration of terms of such a series whenever we attempt to obtain a function from one of its derivatives. Suppose $D_h^n f(x_0 + h) = z$

where h is a variable, x_0 a given value, and z, of course, a function of h. Let $\smallint\!\!^2$ stand for $\smallint\!\!\smallint$, &c., so that $\smallint^n = \smallint\!\!\smallint^{n-1}$.

Then $\qquad D_h^{n-1} f(x_0 + h) = A_1 + \smallint_h z$,

where A_1 is a constant;

$$D_h^{n-2} f(x_0 + h) = A_2 + A_1 h + \smallint_h^2 z,$$

CHAP. IX.] DEVELOPMENT IN SERIES. 119

$$D_h^{n-3}f(x_0+h) = A_3 + A_2 h + \tfrac{1}{2} A_1 h^2 + \int_h^3 z,$$

$$D_h^{n-4}f(x_0+h) = A_4 + A_3 h + \tfrac{1}{2} A_2 h^2 + \frac{1}{2.3} A_1 h^3 + \int_h^4 z,$$

. . . .

$$f(x_0+h) = A_n + A_{n-1} h + \tfrac{1}{2} A_{n-2} h^2 + \frac{1}{2.3} A_{n-3} h^3 + \cdots$$

$$+ \frac{1}{2.3 \cdots (n-1)} A_1 h^{n-1} + \int_h^n z,$$

and we have a set of terms arranged according to the ascending powers of h. Although, by increasing n indefinitely, we can make the second member above a true series, it does not by any means follow that every function can be developed into such a series. In the first place, it may not be possible to increase n indefinitely in the expression above, as the nth derivative of the function may become at last infinite or discontinuous, so that $\int_h^n z$ cannot be dealt with. Next, the series may be a divergent series, and then it could not be equal to the definite value $f(x_0+h)$. But the result is a remarkable one, and suggests the careful investigation of the development of functions in series.

124. Assuming, for the moment, that $f(x_0+h)$ can be developed into a convergent series arranged according to the ascending powers of h, let us see what the coefficients of the series must be. Let

$$f(x_0+h) = A_0 + A_1 h + A_2 h^2 + A_3 h^3 + \cdots + A_n h^n + \cdots$$

The function and the series are both functions of h, and may be differentiated relatively to h.

$$D_h f(x_0+h) = A_1 + 2A_2 h + 3A_3 h^2 + 4A_4 h^3 + \cdots + nA_n h^{n-1} + \cdots$$

We shall find it convenient to adopt the following notation: Let $f'x$ stand for $D_x fx$, $f''x$ for $D_x^2 fx$, $f^{(n)} x$ for $D_x^n fx$. Let

$f'x_0, f^{(n)}x_0$ stand for the results obtained by substituting x_0 for x in $f'x, f^{(n)}x$, where x_0 may be a single term or any complicated function. Let $n!$ (which is to be read n *admiration*) stand for $1 \times 2 \times 3 \times 4 \times \cdots \times n$.

Call $$(x_0 + h) = x,$$

then
$$D_h f(x_0 + h) = D_h fx = D_x fx \, D_h x$$
$$= f'(x_0 + h) \, D_h(x_0 + h) = f'(x_0 + h).$$

In like manner, we could show that

$$D_h^2 f(x_0 + h) = f''(x_0 + h),$$

$$D_h^n(x_0 + h) = f^{(n)}(x_0 + h), \&c.$$

$$f(x_0+h) = A_0 + A_1 h + A_2 h^2 + \cdots + A_n h^n \cdots ;$$
$$f'(x_0+h) = \quad A_1 + 2A_2 h + \cdots + nA_n h^{n-1} + \cdots ;$$
$$f''(x_0+h) = \qquad 2A_2 + \cdots + n(n-1)A_n h^{n-2} + \cdots ;$$
$$f^{(n)}(x_0+h) = \qquad\qquad n!A_n + (n+1)n\cdots 2 A_{n+1} h + \cdots$$

Let $h = 0$ in these equations, and we have

$$fx_0 = A_0, \qquad f'''x_0 = 3!\, A_3,$$

$$f'x_0 = A_1, \qquad f^{\text{IV}}x_0 = 4!\, A_4,$$

$$f''x_0 = 2A_2, \qquad f^{(n)}x_0 = n!\, A_n,$$

hence
$$A_0 = fx_0, \qquad A_3 = \frac{1}{3!} f'''x_0,$$

$$A_1 = f'x_0, \qquad A_4 = \frac{1}{4!} f^{\text{IV}} x_0,$$

$$A_2 = \frac{1}{2} f'' x_0, \qquad A_n = \frac{1}{n!} f^{(n)} x_0,$$

CHAP. IX.] DEVELOPMENT IN SERIES. 121

and $f(x_0+h) = fx_0 + hf'x_0 + \dfrac{h^2}{2!}f''x_0 + \dfrac{h^3}{3!}f'''x_0 + \dfrac{h^4}{4!}f^{IV}x_0$

$$+ \cdots + \dfrac{h^n}{n!}f^{(n)}x_0 + \cdots, \qquad [1]$$

if $f(x_0+h)$ can be developed.

EXAMPLES.

(1) To develop $(a+h)^n$.

Call $\qquad (a+h) = x,$

then $\qquad fx = x^n,$

$$f'x = nx^{n-1},$$

$$f''x = n(n-1)x^{n-2},$$

$$f'''x = n(n-1)(n-2)x^{n-3}, \text{ \&c.}$$

$$fa = a^n,$$

$$f'a = na^{n-1},$$

$$f''a = n(n-1)a^{n-2},$$

$$f'''a = n(n-1)(n-2)a^{n-3}, \text{ \&c.}$$

$$(a+h)^n = a^n + na^{n-1}h + \dfrac{n(n-1)}{2}a^{n-2}h^2$$

$$+ \dfrac{n(n-1)(n-2)}{3!}a^{n-3}h^3 + \cdots,$$

if $(a+h)^n$ *can be developed.*

(2) To develop $\sin h$.

$$\sin h = \sin(0+h).$$

Let $\qquad x = 0 + h.$

$$fx = \sin x, \qquad\qquad f0 = \sin 0 = 0,$$

$$f'x = \cos x, \qquad\qquad f'0 = \cos 0 = 1,$$

$$f''x = -\sin x, \qquad\qquad f''0 = -\sin 0 = 0,$$

$$f'''x = -\cos x, \qquad\qquad f'''0 = -\cos 0 = -1,$$

$$f^{\text{IV}}x = \sin x, \text{ \&c.} \qquad\qquad f^{\text{IV}}0 = \sin 0 = 0, \text{ \&c.}$$

$$\sin(0+h) = 0 + h + 0\cdot\frac{h^2}{2!} - \frac{h^3}{3!} + 0\cdot\frac{h^4}{4!} + \frac{h^5}{5!} + \cdots,$$

$$\sin h = h - \frac{h^3}{3!} + \frac{h^5}{5!} - \frac{h^7}{7!} + \frac{h^9}{9!} + \cdots,$$

if $\sin h$ *can be developed.*

(3) Assuming that $\cos h$ can be developed, determine the series.

125. Let us find what error we are liable to commit if we take $f(x_0 + h)$ equal to $n+1$ terms of the series (Art. 124, [1]). Let R be the difference between $f(x_0 + h)$ and the sum of the first $n+1$ terms; then

$$f(x_0 + h) = fx_0 + hf'x_0 + \frac{h^2}{2!}f''x_0 + \cdots + \frac{h^n}{n!}f^{(n)}x_0 + R,$$

and we want to find the value of R.

Lemma.

126. If a continuous function becomes equal to zero for two different values of the variable, there must be some value of the variable between the two for which the derivative of the function will equal zero.

For, in passing from the first zero value to the second, the function must first increase and then decrease as the variable increases, or first decrease and then increase. If it does the first, the derivative must at some point change from a positive to a negative value; if the second, the derivative must change from a negative to a positive value, and in so doing it must, in either case, pass through the value zero.

127. To determine R.

Let
$$P = R + \frac{h^{n+1}}{(n+1)!},$$

then
$$R = \frac{h^{n+1}}{(n+1)!} P,$$

and
$$f(x_0+h) = fx_0 + hf'x_0 + \frac{h^2}{2!} f''x_0 + \cdots + \frac{h^n}{n!} f^{(n)} x_0 + \frac{h^{n+1}}{(n+1)!} P$$

or
$$f(x_0+h) - fx_0 - hf'x_0 - \frac{h^2}{2!} f''x_0 \cdots - \frac{h^n}{n!} f^{(n)} x_0 - \frac{h^{n+1}}{(n+1)!} P = 0.$$

Call
$$(x_0 + h) = X;$$

then
$$h = X - x_0, \qquad \text{and we have}$$

$$fX - fx_0 - \frac{(X-x_0)}{1!} f'x_0 - \cdots - \frac{(X-x_0)^n}{n!} f^{(n)} x_0$$

$$- \frac{(X-x_0)^{n+1}}{(n+1)!} P = 0. \qquad [1]$$

Form arbitrarily the same function of a variable z that the first member of [1] is of x_0, and call it Fz.

$$Fz = fX - fz - \frac{(X-z)}{1!} f'z - \frac{(X-z)^2}{2!} f''z - \cdots$$

$$- \frac{(X-z)^n}{n!} f^{(n)} z - \frac{(X-z)^{n+1}}{(n+1)!} P.$$

If $z = x_0$, Fz becomes identical with the first member of [1], and therefore $= 0$.

If $\quad z = X, \; Fz = 0$,

since each term disappears from containing a zero factor; and we have succeeded in forming a function of z, which becomes equal to zero for two values, x_0 and X of z. If Fz is continuous, there must be some value of z between x_0 and X for which $F'z = 0$. Differentiating Fz, and remembering that P is constant, we have

$$F'z = 0 - f'z + f'z - \frac{(X-z)}{1!}f''z + \frac{(X-z)}{1!}f''z - \frac{(X-z)^2}{2!}f'''z$$

$$+ \frac{(X-z)^2}{2!}f'''z - \cdots + \frac{(X-z)^{n-1}}{(n-1)!}f^{(n)}z$$

$$- \frac{(X-z)^n}{n!}f^{(n+1)}z + \frac{(X-z)^n}{n!}P.$$

All the terms but the last two destroy one another, and

$$F'z = -\frac{(X-z)^n}{n!}f^{(n+1)}z + \frac{(X-z)^n}{n!}P.$$

But this must be equal to zero for some value of z between x_0 and X. Such a value can be represented by $x_0 + \theta(X - x_0)$ where θ is some positive fraction less than 1, i.e., $0 < \theta < 1$. Substituting this value, we have

$$0 = -\frac{[X - x_0 - \theta(X - x_0)]^n}{n!}f^{(n+1)}[x_0 + \theta(X - x_0)]$$

$$+ \frac{[X - x_0 - \theta(X - x_0)]^n}{n!}P.$$

Whence $\quad P = f^{(n+1)}[x_0 + \theta(X - x_0)].$

$$X - x_0 = h,$$

$$P = f^{(n+1)}(x_0 + \theta h),$$

whence
$$f(x_0+h) = fx_0 + \frac{h}{1!}f'x_0 + \frac{h^2}{2!}f''x_0 + \cdots$$
$$+ \frac{h^n}{n!}f^{(n)}x_0 + \frac{h^{n+1}}{(n+1)!}f^{(n+1)}(x_0+\theta h),$$

where all that we know about θ is that it lies between 0 and 1.

128. The expression for the last term may be obtained in a different form by assuming at the start

$$R = hP$$

instead of
$$R = \frac{h^{n+1}}{(n+1)!}P.$$

Making this assumption, show that
$$f(x_0+h) = fx_0 + hf'x_0 + \frac{h^2}{2!}f''x_0 + \cdots$$
$$+ \frac{h^n}{n!}f^{(n)}x_0 + \frac{h^{n+1}(1-\theta)^n}{n!}f^{(n+1)}(x_0+\theta h).$$

Since in each of these formulas x_0 was any given value, we can represent it in the result just as well by x, and the formulas may be written

$$f(x+h) = fx + hf'x + \frac{h^2}{2!}f''x + \cdots$$
$$+ \frac{h^n}{n!}f^{(n)}x + \frac{h^{n+1}}{(n+1)!}f^{(n+1)}(x+\theta h) ; \qquad [1]$$

$$f(x+h) = fx + hf'x + \frac{h^2}{2!}f''x + \cdots$$
$$+ \frac{h^n}{n!}f^{(n)}x + \frac{h^{n+1}(1-\theta)^n}{n!}f^{(n+1)}(x+\theta h). \qquad [2]$$

and these formulas are known as *Taylor's Theorem*.

Example.

129. To develop $(2+1)^4$.

Let us see what error we are liable to if we stop at the second term.

$$fx = x^4, \qquad f'''x = 24x,$$
$$f'x = 4x^3, \qquad f^{IV}x = 24,$$
$$f''x = 12x^2, \qquad f^{V}x = 0.$$

$$(2+1)^4 = 2^4 + 1.4.2^3 + \frac{1^2}{2!} 12(2+\theta)^2.$$

If $\theta = 0$, the last term is 24. If $\theta = 1$, the last term is 54. Hence, if we stop at the second term, our error lies between 24 and 54. In point of fact, it is 33. Suppose we stop with the third term.

$$(2+1)^4 = 2^4 + 1.4.2^3 + \frac{1^2}{2!} 12.2^2 + \frac{1^3}{3!} 24(2+\theta).$$

If $\theta = 0$, the last term is 8. If $\theta = 1$, the last term is 12, and the error must be between 8 and 12. It is actually 9. Suppose we stop with the fourth term.

$$(2+1)^4 = 2^4 + 1.4.2^3 + \frac{1^2}{2!} 12.2^2 + \frac{1^3}{3!} 24.2 + \frac{1^4}{4!} 24.$$

Here the error is precisely $\dfrac{1^4}{24} 24 = 1$.

Example.

To find $\sin(0+1)$.

Let $n = 7$.

$$fx = \sin x, \qquad f^{IV}x = \sin x,$$
$$f'x = \cos x, \qquad f^{V}x = \cos x,$$
$$f''x = -\sin x, \qquad f^{VI}x = -\sin x, \&c.$$
$$f'''x = -\cos x,$$

$$\sin(0+1) = 1 - \frac{1}{3!} + \frac{1}{5!} - \frac{1}{7!} + \frac{1}{8!} \sin\theta.$$

If $\theta = 0$,
$$\frac{\sin\theta}{8!} = 0.$$

If $\theta = 1$,
$$\frac{\sin\theta}{8!} = \frac{\sin 1}{40320}.$$

$\frac{4211}{5040}$ is within $\frac{1}{40320}$ of the true value of sin 1.

If in any development the general expression for the error decreases indefinitely as we increase n, *it follows that, as the number of terms of the series is indefinitely increased, the sum will approach as its limit the value of the function, which is therefore equal to a series of the form obtained, and is said to be developable.*

130. Let us consider some examples.
To develop $\log(1+x)$.
Let $z = (1+x)$.

$$fz = \log z,$$
$$f'z = z^{-1},$$
$$f''z = -z^{-2},$$
$$f'''z = 2z^{-3},$$
$$f^{IV}z = -3!\, z^{-4},$$
$$f^{(n)}z = (-1)^{n-1}(n-1)!\, z^{-n},$$
$$f^{(n+1)}z = (-1)^n n!\, z^{-n-1},$$
$$f(1) = 0,$$
$$f'(1) = 1,$$
$$f''(1) = -1,$$
$$f'''(1) = 2,$$

$$f^{iv}(1) = -3!,$$

$$f^{(n)}(1) = (-1)^{n-1}(n-1)!,$$

$$f^{(n+1)}(1+\theta x) = (-1)^n n!(1+\theta x)^{-n-1}.$$

By Taylor's Theorem,

$$\log(1+x) = x - \frac{x^2}{2} + \frac{x^3}{3} - \frac{x^4}{4} + \cdots + (-1)^{n-1}\frac{x^n}{n}$$

$$+ \frac{(-1)^n x^{n+1}}{n+1}(1+\theta x)^{-n-1}.$$

The ratio of the nth term to the term before it is $\dfrac{(-1)(n-1)x}{n}$, or $-\left(1-\dfrac{1}{n}\right)x$. If x is greater than 1 in absolute value, $\left(1-\dfrac{1}{n}\right)x$ will eventually become and remain greater than unity as n increases, and the series

$$x - \frac{x^2}{2} + \frac{x^3}{3} - \frac{x^4}{4} + \cdots$$

is divergent and cannot be equal to $\log(1+x)$. So we need only investigate the expression for the error for the values of x between $+1$ and -1. Suppose x is positive, and less than 1. Then $\dfrac{x^{n+1}}{n+1}(1+\theta x)^{-n-1}$ approaches zero as its limit as n increases indefinitely, for it may be thrown into the form $\dfrac{1}{n+1}\left(\dfrac{x}{1+\theta x}\right)^{n+1}$. Since $x < 1$, $\dfrac{x}{1+\theta x} < 1$. $\left(\dfrac{x}{1+\theta x}\right)^{n+1}$ has zero for its limit as n increases indefinitely; as has also the factor $\dfrac{1}{n+1}$. Hence, for values of x between 0 and 1, $\log(1+x)$ is developable, and is equal to the series

$$x - \frac{x^2}{2} + \frac{x^3}{3} - \frac{x^4}{4} + \cdots$$

This is true even where $x=1$, for it is easily seen that, in that case also, $\dfrac{1}{n+1}\left(\dfrac{x}{1+\theta x}\right)^{n+1}$ approaches the limit zero as n increases. If x is between 0 and -1, the second form of the error, Art. 128, [2], is most convenient for our purpose. Let $x = -x'$, so that x' is positive and less than 1. Then our function is $\log(1-x')$, and the series becomes

$$-x' - \frac{x'^2}{2} - \frac{x'^3}{3} - \cdots - \frac{x'^n}{n} - \frac{x'^{n+1}(1-\theta)^n}{(1-\theta x')^{n+1}}.$$

$$\frac{x'^{n+1}(1-\theta)^n}{(1-\theta x')^{n+1}} = \frac{x'}{1-\theta x'}\left(\frac{x'-\theta x'}{1-\theta x'}\right)^n;$$

where $\dfrac{x'-\theta x'}{1-\theta x'}$ is less than 1;

hence
$$\lim_{n=\infty}\left(\frac{x'-\theta x'}{1-\theta x'}\right)^n = 0,$$

and as $\dfrac{x'}{1-\theta x'}$ is a finite value, the expression for the error decreases indefinitely as n increases, and the function is equal to the series. Our expansion

$$\log(1+x) = x - \frac{x^2}{2} + \frac{x^3}{3} - \frac{x^4}{4} + \cdots$$

holds, then, for values of x between 1 and -1.

The Binomial Theorem.

131. To develop $(1+x)^m$.

Let
$$z = 1 + x,$$
$$fz = z^m,$$
$$f'z = mz^{m-1},$$
$$f''z = m(m-1)z^{m-2},$$
$$f'''z = m(m-1)(m-2)z^{m-3},$$
$$f^{(n)}z = m(m-1)\cdots(m-n+1)z^{m-n},$$
$$f^{(n+1)}z = m(m-1)\cdots(m-n)z^{m-n-1},$$

$$f(1)=1,$$

$$f'(1)=m.$$

$$f''(1)=m(m-1),$$

$$f'''(1)=m(m-1)(m-2),$$

$$f^{(n)}(1)=m(m-1)\cdots(m-n+\,),$$

$$f^{(n+1)}(1+\theta x)=m(m-1)\cdots(m-n)(1+\theta x)^{m-n-1}.$$

By Taylor's Theorem,

$$(1+x)^m = 1 + mx + \frac{m(m-1)x^2}{2!} + \frac{m(m-1)(m-2)x^3}{3!} + \cdots$$

If m is a positive whole number,

$$f^{(m)}z = m!,$$

$$f^{(m+1)}z = 0,$$

and all succeeding derivatives are 0, so in that case $(1+x)^m$ is equal to the sum of a finite number of terms, namely $(m+1)$ terms. If m is negative or fractional, however, this is not the case. Let us see whether $(1+x)^m$ is then developable. The ratio of the general term of the series to the one before it is $\frac{m-n+1}{n}x$ or $\left(\frac{m+1}{n}-1\right)x$. If x is numerically greater than 1, this ratio will eventually become and remain greater than 1 in absolute value as n increases, and the series is divergent and cannot be equal to the function. Hence we need examine the value of the error only for values of x between 1 and -1. The expression for the remainder after $n+1$ terms is

$$\frac{m(m-1)\cdots(m-n)}{(n+1)!}x^{n+1}(1+\theta x)^{m-n-1},$$

which may be thrown into the form

$$\left[\frac{m(m-1)\cdots(m-n)}{(n+1)!}x^{n+1}\right]\frac{1}{(1+\theta x)^{n+1-m}}.$$

As n increases, the limit approached by $\dfrac{1}{(1+\theta x)^{n+1-m}}$ is not greater than 1. Increasing n by unity multiplies the quantity in parenthesis by $\dfrac{m-n-1}{n+2}\, x$, which may be written

$$\left(\dfrac{m-1}{n+2} - \dfrac{n}{n+2}\right)x\,;$$

and by taking n sufficiently large, this multiplier may be brought as near as we please to the value $-x$. If x lies between 0 and 1, $-x$ is numerically less than 1 ; and as n increases indefinitely, we multiply our parenthesis by an indefinite number of factors, each less than 1, and so decrease the product indefinitely. Therefore, for values of x between 0 and 1, the expression for the error approaches zero as its limit as n increases indefinitely, and $(1+x)^m$ is equal to the series

$$1 + mx + \dfrac{m(m-1)}{2!}x^2 + \dfrac{m(m-1)(m-2)}{3!}x^3 + \cdots \quad [1]$$

Example.

Show, by considering the second form for the error, Art. 128, [2], that for values of x between 0 and -1, $(1+x)^m$ is developable.

The *Binomial Theorem* follows easily from the development of $(1+x)^m$.

$$(x+h)^m = x^m\left(1 + \dfrac{h}{x}\right)^m;$$

and if h is less than x in absolute value, we have

$$(x+h)^m = x^m + mx^{m-1}h + \dfrac{m(m-1)}{2!}x^{m-2}h^2$$
$$+ \dfrac{m(m-1)(m-2)}{3!}x^{m-3}h^3 + \cdots, \quad [2]$$

no matter what the value of m.

Of course, if h is greater than x, we can write $(x+h)^m$ in the form $h^m\left(1+\dfrac{x}{h}\right)^m$, and shall then get as a true development,

$$(h+x)^m = h^m + mh^{m-1}x + \cdots$$

Maclaurin's Theorem.

132. If, in Art. 128, [1] and [2], we let $x = 0$, we get

$$fh = f(0) + hf'(0) + \frac{h^2}{2!}f''(0) + \frac{h^3}{3!}f'''(0) + \cdots$$

$$+ \frac{h^n}{n!}f^{(n)}(0) + \frac{h^{n+1}}{(n+1)!}f^{(n+1)}\theta h,$$

$$fh = f(0) + hf'(0) + \frac{h^2}{2!}f''(0) + \frac{h^3}{3!}f'''(0) + \cdots$$

$$+ \frac{h^n}{n!}f^{(n)}(0) + \frac{h^{n+1}(1-\theta)^n}{n!}f^{(n+1)}\theta h.$$

It does not matter what letter we use for the variable in these formulas. Change h to x, and

$$fx = f(0) + xf'(0) + \frac{x^2}{2!}f''(0) + \cdots + \frac{x^n}{n!}f^{(n)}(0)$$

$$+ \frac{x^{n+1}}{(n+1)!}f^{(n+1)}\theta x. \qquad [1]$$

$$fx = f(0) + xf'(0) + \frac{x^2}{2!}f''(0) + \cdots + \frac{x^n}{n!}f^{(n)}(0)$$

$$+ \frac{x^{n+1}(1-\theta)^n}{n!}f^{(n+1)}\theta x. \qquad [2]$$

These results are called *Maclaurin's Theorem*, and they enable us to develop a function in a series arranged according to the ascending powers of the variable.

133. To develop a^x.

$$fx = a^x, \qquad f(0) = a^0 = 1.$$
$$f'x = a^x \log a, \qquad f'(0) = \log a,$$
$$f''x = a^x (\log a)^2, \qquad f''(0) = (\log a)^2,$$
$$f^{(n)}x = a^x (\log a)^n, \qquad f^{(n)}(0) = (\log a)^n,$$
$$f^{(n+1)}x = a^x (\log a)^{n+1}, \qquad f^{(n+1)}\theta x = a^{\theta x}(\log a)^{n+1}.$$

By Art. 132, [1],

$$a^x = 1 + x \log a + \frac{x^2}{2!}(\log a)^2 + \frac{x^3}{3!}(\log a)^3 + \cdots + \frac{x^n}{n!}(\log a)^n$$

$$+ \frac{x^{n+1}}{(n+1)!}(\log a)^{n+1} a^{\theta x}.$$

$$\frac{x^{n+1}(\log a)^{n+1}}{(n+1)!} a^{\theta x} = a^{\theta x} \cdot \frac{x \log a}{1} \cdot \frac{x \log a}{2} \cdot \frac{x \log a}{3} \cdots \frac{x \log a}{n} \cdot \frac{x \log a}{n+1}.$$

No matter what value x may have, after n has attained a certain value in its increase, some of the factors of this product will approach the limit zero, and the whole product will therefore have zero for its limit as n increases indefinitely, and

$$a^x = 1 + x \log a + \frac{x^2}{2!}(\log a)^2 + \frac{x^3}{3!}(\log a)^3 + \cdots \qquad [1]$$

for all values of x. If $a = e$, $\log a = 1$,

and

$$e^x = 1 + \frac{x}{1} + \frac{x^2}{2!} + \frac{x^3}{3!} + \frac{x^4}{4!} + \cdots \qquad [2]$$

Let $x = 1$, and [2] becomes

$$e = 1 + \frac{1}{1} + \frac{1}{2!} + \frac{1}{3!} + \frac{1}{4!} + \cdots ; \qquad [3]$$

a result already established in Arts. 61 and 62.

134. We can now test the accuracy of the provisional developments of sine and cosine given in Art. 124, (2) and (3). By Art. 132, [1],

$$\sin x = x - \frac{x^3}{3!} + \frac{x^5}{5!} - \frac{x^7}{7!} + \cdots + R,$$

where $$R = \frac{x^{n+1}}{(n+1)!} f^{(n+1)} \theta x = \pm \frac{x^{n+1}}{(n+1)!} \sin \theta x$$

or $$\pm \frac{x^{n+1}}{(n+1)!} \cos \theta x.$$

In either case, one factor $\sin \theta x$ or $\cos \theta x$ is between 1 and -1, and the other approaches zero as n increases indefinitely; therefore,

$$\sin x = x - \frac{x^3}{3!} + \frac{x^5}{5!} - \frac{x^7}{7!} + \frac{x^9}{9!} - \cdots$$

Example.

Prove that $\cos x = 1 - \dfrac{x^2}{2!} + \dfrac{x^4}{4!} - \dfrac{x^6}{6!} + \dfrac{x^8}{8!} - \cdots$

135. By the aid of the Binomial Theorem, $\tan^{-1} x$ and $\sin^{-1} x$ can be very easily developed.

$$D_x \tan^{-1} x = \frac{1}{1+x^2} = (1+x^2)^{-1}. \quad \text{(Art. 71, Ex.)}$$

For values of x less than 1, $(1+x^2)^{-1}$ can be developed by Art. 131, [2], $(1+x^2)^{-1} = 1 - x^2 + x^4 - x^6 + x^8 - \cdots$

Integrate both members.

$$\tan^{-1} x = C + x - \frac{x^3}{3} + \frac{x^5}{5} - \frac{x^7}{7} + \frac{x^9}{9} - \cdots$$

To determine our arbitrary constant C, let

$$x = 0;$$

then
$$\tan^{-1} 0 = C + 0 - \frac{0}{3} \ldots ,$$

and
$$C = 0.$$

$$\tan^{-1} x = x - \frac{x^3}{3} + \frac{x^5}{5} - \frac{x^7}{7} + \frac{x^9}{9} - \ldots \qquad [1]$$

when x is less than 1; that is, when $\tan^{-1} x$ is less than $\frac{\pi}{4}$.

$$D_x \sin^{-1} x = \frac{1}{\sqrt{(1-x^2)}} = (1-x^2)^{-\frac{1}{2}}, \quad \text{by Art. 71.}$$

For values of x less than 1, $(1-x^2)^{-\frac{1}{2}}$ can be developed by Art. 131, [2].

$$(1-x^2)^{-\frac{1}{2}} = 1 + \frac{1}{2} x^2 + \frac{1.3}{2.4} x^4 + \frac{1.3.5}{2.4.6} x^6 + \frac{1.3.5.7}{2.4.6.8} x^8 + \ldots$$

Integrating
$$\sin^{-1} x = C + x + \frac{1}{2} \frac{x^3}{3} + \frac{1.3}{2.4} \frac{x^5}{5} + \frac{1.3.5}{2.4.6} \frac{x^7}{7} + \ldots$$

When
$$x = 0,$$

$$\sin^{-1} x = 0 \text{ and } C = 0.$$

$$\sin^{-1} x = x + \frac{1}{2} \frac{x^3}{3} + \frac{1.3}{2.4} \frac{x^5}{5} + \frac{1.3.5}{2.4.6} \frac{x^7}{7} + \ldots \qquad [2]$$

Examples.

(1) Show that $\sin(x+h)$ is equal to the series

$$\sin x + \frac{h}{1!} \cos x - \frac{h^2}{2!} \sin x - \frac{h^3}{3!} \cos x + \frac{h^4}{4!} \sin x + \ldots$$

(2) Show that
$$e^{mx} = 1 + \frac{mx}{1!} + \frac{m^2 x^2}{2!} + \frac{m^3 x^3}{3!} + \ldots$$

136. Although the strict proof that any given function is equal to the series obtained by Taylor's Theorem requires the investigation of the *remainder after* n+1 *terms*, it is often convenient to obtain terms of the series in cases where the expression for the remainder is too complicated to admit of the usual examination. When such a series is employed, it is to be remembered that it is equal to the function in question only provided that the function is developable. Sometimes the possibility of development can be established by other considerations, and sometimes in rough work no attempt is made to fill out the proof of the assumed equality.

Examples.

(1) Develop $\dfrac{x}{1+x} + \log(1+x)$.

Ans. $2x - \dfrac{3}{2}x^2 + \dfrac{4}{3}x^3 - \dfrac{5}{4}x^4 + \dfrac{6}{5}x^5 + \cdots$

(2) Obtain 4 terms of the development of $\log(1+e^x)$.

Ans. $\log 2 + \dfrac{x}{2} + \dfrac{x^2}{2^3} - \dfrac{x^4}{2^3 . 4\,!}$.

137. In the work of successive differentiation required in applying Taylor's Theorem, a good deal of labor can often be saved by making use of *Leibnitz's Theorem for the Derivatives of a Product*. Let y and z be functions of x. Represent $D_x y, D_x^2 y, \cdots D_x^n y$ by $y', y'', \cdots y^{(n)}$ and $D_x z, D_x^2 z, \cdots D_x^n z$ by $z', z'', \cdots z^{(n)}$.

$$D_x(yz) = y'z + yz',$$

$$D_x^2(yz) = y''z + 2y'z' + yz'',$$

$$D_x^3(yz) = y'''z + 3y''z' + 3y'z'' + yz''',$$

$$D_x^4(yz) = y^{\text{IV}}z + 4y'''z' + 6y''z'' + 4y'z''' + yz^{\text{IV}}.$$

Examining these results, we see that the coefficients of the terms in the successive derivatives are the same as in the corresponding powers of a binomial, and that the accents follow the same law as the exponents in the powers of a binomial. Following the same analogy, we should have

$$D_x^n(yz) = y^{(n)}z + ny^{(n-1)}z' + \frac{n(n-1)}{2!}y^{(n-2)}z''$$
$$+ \frac{n(n-1)(n-2)}{3!}y^{(n-3)}z''' + \ldots$$

Assuming for the moment the truth of this equation, let us differentiate both members. We obtain

$$D_x^{n+1}(yz) = y^{(n+1)}z + (n+1)y^{(n)}z' + \frac{(n+1)n}{2!}y^{n-1}z''$$
$$+ \frac{(n+1)n(n-1)}{3!}y^{(n-2)}z''' + \ldots;$$

but this is precisely what we should expect for the $(n+1)$st derivative from the observed analogy. Hence, if our rule holds for the nth derivative, it holds for the $(n+1)$st; but we have seen that it holds for the 4th, therefore it holds for the 5th, and therefore for the 6th, and so on; and it is in consequence universally true. This rule is called *Leibnitz's Theorem*, and is formulated as follows:

$$D_x^n(yz) = y^{(n)}z + ny^{(n-1)}z' + \frac{n(n-1)}{2!}y^{(n-2)}z''$$
$$+ \frac{n(n-1)(n-2)}{3!}y^{(n-3)}z'''\ldots \qquad [1]$$

138. Assuming that $\tan x$ can be developed, let us obtain a few terms of the series. Here

$fx = \tan x = y,$

$f'x = y' = \sec^2 x,$

$f''x = y'' = 2\sec^2 x \tan x = 2y'y,$

$f'''x = y''' = 2(y''y + y'y'),$

$f^{\text{IV}}x = y^{\text{IV}} = 2(y'''y + 2y''y' + y'y''),$

$$f^v x = y^v = 2(y^{iv}y + 3y'''y' + 3y''y'' + y'y'''),$$

$$f^{vi} x = y^{vi} = 2(y^v y + 4y^{iv}y' + 6y'''y'' + 4y''y''' + y'y^{iv}),$$

$$f^{vii} x = y^{vii} = 2(y^{vi}y + 5y^v y' + 10y^{iv}y'' + 10y'''y''' + 5y''y^{iv} + y'y^v),$$

&c., by Leibnitz's Theorem.

When $x = 0$,

$y = 0$, $y^{iv} = 0$,

$y' = 1$, $y^v = 16$,

$y'' = 0$, $y^{vi} = 0$,

$y''' = 2$, $y^{vii} = 272$.

By Maclaurin's Theorem

$$\tan x = x + \frac{2}{3!}x^3 + \frac{16}{5!}x^5 + \frac{272}{7!}x^7 + \dots$$

EXAMPLE.

Assuming that $\sec x$ can be developed, show that

$$\sec x = 1 + \frac{x^2}{2!} + \frac{5x^4}{4!} + \frac{61x^6}{6!} + \dots$$

Indeterminate Forms.

139. The subject of indeterminate forms is readily dealt with by the aid of Taylor's Theorem. Take the form $\frac{0}{0}$. Suppose fx and Fx are functions of x, continuous for values of x near the particular value a, and fa and Fa are both equal to zero, to find the *true value* (*vide* Art. 34) of $\frac{fx}{Fx}$ when $x = a$.

Call $x - a = h$, then $x = a + h$,

and we can develop fx and Fx by Taylor's Theorem.

DEVELOPMENT IN SERIES.

$$fx = f(a+h) = fa + hf'(a+\theta h)$$

where θ is some number between zero and 1.

$$Fx = F(a+h) = Fa + hF'(a+\theta' h)$$

where $0 < \theta' < 1$.

$$\frac{fx}{Fx} = \frac{hf'(a+\theta h)}{hF'(a+\theta' h)} = \frac{f'(a+\theta h)}{F'(a+\theta' h)},$$

since $fa = 0$ and $Fa = 0$.

As x approaches a, h approaches zero; hence θh and $\theta' h$ approach zero as their limit; consequently the limit approached by $\frac{fx}{Fx}$ as x approaches a, is $\frac{f'a}{F'a}$, which, by Art. 34, is the *true value* of $\frac{fa}{Fa}$. If $fa = 0$, $Fa = 0$, $f'a = 0$, and $F'a = 0$, it will be necessary to carry the development one step farther.

$$fx = f(a+h) = fa + hf'a + \frac{h^2}{2!}f''(a+\theta h) = \frac{h^2}{2!}f''(a+\theta h),$$

$$Fx = F(a+h) = Fa + hF'a + \frac{h^2}{2!}F''(a+\theta' h) = \frac{h^2}{2!}F''(a+\theta' h),$$

and
$$\frac{fx}{Fx} = \frac{f''(a+\theta h)}{F''(a+\theta' h)},$$

which approaches $\frac{f''a}{F''a}$ as its limit as x approaches a.

EXAMPLE.

Show that, if fa, Fa, $f'a$, $F'a$, $f''a$, $F''a$, &c., $f^{(n-1)}a$, and $F^{(n-1)}a$ all equal zero, the true value of $\frac{fx}{Fx}$ when $x = a$ is $\frac{f^{(n)}a}{F^{(n)}a}$.

140. The reasoning of the last section does not apply when

$a = \infty$, as then $f(a+h)$ cannot be developed by Taylor's Theorem.

To find the true value of $\dfrac{fx}{Fx}$ when $x = \infty$, supposing that

$$fx = 0 \text{ and } Fx = 0 \text{ when } x = \infty.$$

Let
$$y = \frac{1}{x},$$

then
$$fx = f\frac{1}{y} \text{ and } Fx = F\frac{1}{y},$$

and $\dfrac{f\dfrac{1}{y}}{F\dfrac{1}{y}}$ assumes the form $\dfrac{0}{0}$ when $y = 0$,

and its true value for $y = 0$ will be $\dfrac{\left[D_y f\dfrac{1}{y}\right]_{y=0}}{\left[D_y F\dfrac{1}{y}\right]_{y=0}}.$

$$D_y f\frac{1}{y} = f'\frac{1}{y} \, D_y \frac{1}{y} = -\frac{1}{y^2} f'\frac{1}{y},$$

$$D_y F\frac{1}{y} = -\frac{1}{y^2} F''\frac{1}{y}.$$

But the value of $\dfrac{D_y f\dfrac{1}{y}}{D_y F\dfrac{1}{y}}$ when $y = 0$

is the value it approaches as y approaches 0.

$$\frac{D_y f\dfrac{1}{y}}{D_y F\dfrac{1}{y}} = \frac{-\dfrac{1}{y^2} f'\dfrac{1}{y}}{-\dfrac{1}{y^2} F'\dfrac{1}{y}} = \frac{f'\dfrac{1}{y}}{F'\dfrac{1}{y}} = \frac{f'x}{F'x};$$

but when $y = 0, \; x = \infty$;

hence the true value of $\dfrac{fx}{Fx}$ when $x = \infty$

is the value of $\dfrac{f'x}{F'x}$ when $x = \infty$;

and the method of the last section holds, no matter what the value of a.

141. It was shown in Art. 35, that the form $\frac{\infty}{\infty}$ could always be reduced to $\frac{0}{0}$ and treated as above. Let us consider a general example. Suppose $fa = \infty$ and $Fa = \infty$, required the true value of $\frac{fx}{Fx}$ when $x = a$.

$$\frac{fx}{Fx} = \frac{\frac{1}{Fx}}{\frac{1}{fx}} = \frac{0}{0} \text{ when } x = a.$$

Differentiate numerator and denominator.

$$D_x \frac{1}{fx} = -\frac{1}{(fx)^2} f'x.$$

$$D_x \frac{1}{Fx} = -\frac{1}{(Fx)^2} F'x.$$

Hence we have, when $x = a$,

$$\frac{fx}{Fx} = \frac{-\frac{1}{(Fx)^2} F'x}{-\frac{1}{(fx)^2} f'x} = \left(\frac{fx}{Fx}\right)^2 \frac{F'x}{f'x}.$$

$$1 = \frac{fx}{Fx} \cdot \frac{F'x}{f'x}$$

and

$$\frac{fx}{Fx} = \frac{f'x}{F'x},$$

the value required. Therefore the form $\frac{\infty}{\infty}$ can be treated directly by the same method as the form $\frac{0}{0}$. In dividing both members of the equation $\qquad \dfrac{fx}{Fx} = \left(\dfrac{fx}{Fx}\right)^2 \dfrac{F'x}{f'x} \qquad$ by $\dfrac{fx}{Fx}$,

we have assumed that the true value of $\dfrac{fx}{Fx}$, when
$$x = a$$
is neither 0 nor ∞. Suppose the limit approached by $\dfrac{fx}{Fx}$ as x approaches a is 0, and that fx and Fx increase indefinitely. Form the function $\dfrac{fx + Fx}{Fx}$.

Its true value when $\qquad x = a, \qquad$ is, of course, 1;

but when $\qquad x = a, \qquad$ it assumes the form $\dfrac{\infty}{\infty}$;

hence its true value when $\quad x = a$,

must be the limit approached by $\dfrac{f'x + F'x}{F'x}$ as x approaches a,

which is $\qquad 1 + \underset{x \doteq a}{\text{limit}} \left[\dfrac{f'x}{F'x} \right].$

Therefore, $\quad \underset{x \doteq a}{\text{limit}} \left[\dfrac{f'x}{F'x} \right] = 0 = \underset{x \doteq a}{\text{limit}} \left[\dfrac{fx}{Fx} \right] \quad$ by hypothesis.

If the true value of $\dfrac{fx}{Fx}$ when $x = a$

is infinite, of course the true value of its reciprocal $\dfrac{Fx}{fx}$ will be zero, and will equal $\left[\dfrac{F'x}{f'x} \right]_{x=a}$;

hence $\qquad \left[\dfrac{f'x}{F'x} \right]_{x=a} = \infty = \left[\dfrac{fx}{Fx} \right]_{x=a},$

and the method of determining the form $\dfrac{\infty}{\infty}$, established at the beginning of this section, is of universal application.

142. The forms ∞^0, 1^∞, 0^0, can all be reduced to one of the forms already discussed, if we make use of logarithms. *It is to be observed, that these forms to be indeterminate must all occur as limiting forms of a function of two functions;* and, in order that the forms may admit of being determined, *the two functions must depend upon the same variable.*

Let
$$u = (Fx)^{fx}.$$

Suppose, when $x = a$,
$$Fx = \infty \text{ and } fx = 0;$$

to find the true value of u when $x = a$.

$$\log u = fx \cdot \log Fx = 0 \times \infty \text{ when } x = a,$$

and may be determined by the method of Art. 35.

Examples.

(1) Show that 1^∞, 0^0, can be made to depend upon the forms $\infty \times 0$ and $0(-\infty)$.

(2) Obtain a method for dealing with the form $\infty - \infty$.

Find the true value of the following functions:—

(3) $\dfrac{\log x}{x-1}$ when $x = 1$. Ans. 1.

(4) $\dfrac{e^x - e^{-x}}{\sin x}$ " $x = 0$. Ans. 2.

(5) $\dfrac{x - \sin^{-1} x}{\sin^3 x}$ " $x = 0$. Ans. $-\tfrac{1}{6}$.

(6) $\dfrac{1}{\log x} - \dfrac{x}{\log x}$ " $x = 1$. Ans. -1.

(7) $\dfrac{e^x - 2\cos x + e^{-x}}{x \sin x}$ " $x = 0$. Ans. 2.

(8) $x\tan x - \dfrac{\pi}{2}\sec x$ when $x = \dfrac{\pi}{2}$. *Ans.* -1.

(9) $2^x \sin \dfrac{a}{2^x}$ " $x = \infty$. *Ans.* a.

(10) $(a^{\frac{1}{x}} - 1)x$ " $x = \infty$. *Ans.* $\log a$.

(11) $\left(\dfrac{a}{x} + 1\right)^x$ " $x = \infty$. *Ans.* e^a.

(12) $\left(1 + \dfrac{1}{x^2}\right)^x$ " $x = \infty$. *Ans.* 1.

(13) $\left(\dfrac{\tan x}{x}\right)^{\frac{1}{x}}$ " $x = 0$. *Ans.* 1.

(14) $\left(\dfrac{\tan x}{x}\right)^{\frac{1}{x^2}}$ " $x = 0$. *Ans.* $e^{\frac{1}{3}}$.

(15) $\left(\dfrac{\tan x}{x}\right)^{\frac{1}{x^3}}$ " $x = 0$. *Ans.* ∞.

(16) $\sin x^{\tan x}$ " $x = \dfrac{\pi}{2}$. *Ans.* 1.

Maxima and Minima.

143. Taylor's Theorem enables us to give a very simple and complete treatment of the subject of maxima and minima of a a single variable.

Let fx be a function of x, finite and continuous for values of x near the particular value a.

Call $\qquad x = a + h.$

$$f(a+h) = fa + hf'a + \dfrac{h^2}{2!}f''(a + \theta h).$$

$$f(a+h) - fa = hf'a + \dfrac{h^2}{2!}f''(a + \theta h).$$

In order that fa should be either a maximum or a minimum,

$$f(a+h) - fa$$

must have the same sign for small values of h whether h is positive or negative. If this sign is minus, fa is a maximum value of fx; if plus, a minimum value (*vide* Art. 39).

If $f'a$ does not equal zero, we can take a value of h so small, that, for it and all smaller values,

$$\frac{h}{2!} f''(a + \theta h)$$

shall be less than $f'a$. The sign of

$$hf'a + \frac{h^2}{2!} f''(a + \theta h)$$

will then, as h approaches zero, ultimately become and remain the same as the sign of $hf'a$; but the sign of $hf'a$ changes with the sign of h, so that fa can be neither a maximum nor a minimum.

144. Suppose $\qquad f'a = 0,$

then $\qquad f(a+h) - fa = \frac{h^2}{2!} f''a + \frac{h^3}{3!} f'''(a + \theta h)$

$$= h^2 \left[\frac{f''a}{2!} + \frac{h}{3!} f'''(a + \theta h) \right];$$

as h approaches zero $\quad \dfrac{h}{3!} f'''(a + \theta h)$

will, in the end, become and remain less than $\dfrac{f''a}{2!}$ and the quantity in parenthesis will have the same sign as $f''a$. As h^2 is necessarily positive for all values of h

$$f(a+h) - fa$$

will then be negative for small positive and negative values of h, if $f'''a$ is negative, and will be a *maximum*; if $f'a$ is positive, fa will be a *minimum*.

145. It can be easily established by an extension of the reasoning of the last section, that, *if the first derivative that does not vanish when* $x = a$ *is of odd order*, fa *is neither a maximum nor a minimum; that, if it is of even order and negative*, fa *is a maximum; if of even order and positive*, fa *is a minimum*.

Examples.

(1) A body moves with different uniform velocities in two different media separated by a plane, required the path of quickest passage from a given point in the first medium to a given point of the second. It is easily seen that the required path will lie in a plane passing through the two given points and perpendicular to the plane separating the two media.

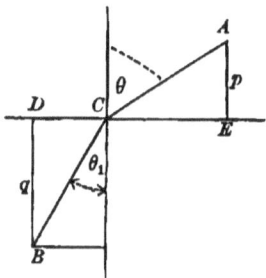

Let ACB represent any such path from A to B. Draw a normal to the plane at C and the perpendiculars p and q. Call

$$DE = c,$$

and let v_1 and v_2 be the velocities in the first and second media respectively.

$$AC = p \sec \theta,$$

$$CE = p \tan \theta,$$

$$BC = q \sec \theta_1,$$

$$DC = q \tan \theta_1,$$

$$p \tan \theta + q \tan \theta_1 = c.$$

$$\frac{AC}{v_1} = \frac{p \sec \theta}{v_1}$$

is the time required to pass from A to C;

$$\frac{BC}{v_2} = \frac{q \sec \theta_1}{v_2}$$

is the time required to pass from C to B;

$$t = \frac{p \sec \theta}{v_1} + \frac{q \sec \theta_1}{v_2}$$

is the function we wish to make a minimum. θ and θ_1 are the only variables in t, and they are connected by the relation

$$p \tan \theta + q \tan \theta_1 = c.$$

$$D_\theta t = \frac{p}{v_1} \sec \theta \tan \theta + \frac{q}{v_2} \sec \theta_1 \tan \theta_1 \, D_\theta \theta_1.$$

Differentiate $\quad p \tan \theta + q \tan \theta_1 = c.$

$$p \sec^2 \theta + q \sec^2 \theta_1 \, D_\theta \theta_1 = 0,$$

$$D_\theta \theta_1 = -\frac{p \sec^2 \theta}{q \sec^2 \theta_1},$$

$$D_\theta t = \frac{p}{v_1} \sec \theta \tan \theta - \frac{q}{v_2} \sec \theta_1 \tan \theta_1 \frac{p \sec^2 \theta}{q \sec^2 \theta_1}.$$

$D_\theta t$ must equal zero in order that t may be a minimum. Express everything in terms of sine and cosine.

$$\frac{p}{v_1} \frac{\sin \theta}{\cos^2 \theta} - \frac{q}{v_2} \frac{\sin \theta_1}{\cos^2 \theta_1} \frac{p}{q} \frac{\cos^2 \theta_1}{\cos^2 \theta} = 0,$$

$$\frac{\sin\theta}{v_1} = \frac{\sin\theta_1}{v_2},$$

$$\frac{\sin\theta}{\sin\theta_1} = \frac{v_1}{v_2}.$$

By taking $D_\theta^2 t$ and substituting

$$\frac{\sin\theta}{\sin\theta_1} = \frac{v_1}{v_2},$$

we should obtain a positive result; so that this relation between the angles gives the path of quickest passage required. This result is the well-known law of the refraction of light, and our solution establishes the fact that a ray of light, in passing from a point in one medium to a point in another, takes the course that enables it to accomplish its journey in the least possible time.

(2) What value of x will make $\sin^3 x \cos x$ a maximum?

$$\textit{Ans. } x = \frac{\pi}{3}.$$

(3) What value of x will make $\sin x(1+\cos x)$ a maximum?

$$\textit{Ans. } x = \frac{\pi}{3}.$$

(4) Show that $x^{\frac{1}{x}}$ is a maximum when $x = e$.

(5) A statue a feet high stands on a column b feet high; how far from the foot of the column must an observer stand that the statue may subtend the greatest possible visual angle?

$$\textit{Ans. } \sqrt{b(a+b)} \text{ feet.}$$

(6) Required the shortest distance from the point (x_0, y_0) to the line $Ax + By + C = 0$.

$$\textit{Ans. } \frac{Ax_0 + By_0 + C}{\sqrt{(A^2 + B^2)}}.$$

CHAPTER X.

INFINITESIMALS.

146. An *infinitesimal* or *infinitely small* quantity is *a variable which is supposed to decrease indefinitely;* in other words, it is *a variable which approaches the limit zero.*

What we have called the *increment* of a variable has, in every case considered, been such a quantity; and what we have called a *derivative* has been the limit of the ratio of infinitesimal increments of function and variable.

147. When we have occasion to consider several infinitesimals connected by some law, we choose arbitrarily some one as the *principal infinitesimal.*

Any infinitesimal such that the limit of its ratio to the principal infinitesimal is finite, is called an *infinitesimal of the first order.*

An infinitesimal such that the limit of its ratio to the square of the principal infinitesimal is finite, is called an *infinitesimal of the second order.*

An infinitesimal such that the limit of its ratio to the nth power of the principal infinitesimal is finite, is called an *infinitesimal of the nth order.*

Let a represent the principal infinitesimal, and a_1 any infinitesimal of the first order, a_2 of the second order, a_n of the nth order. Then, by our definition,

$$\text{limit } \frac{a_1}{a} = K,$$

K being a finite quantity.

$$\frac{a_1}{a} = K + \varepsilon,$$

where ε is an infinitesimal (Art. 7),

$$a_1 = a(K + \varepsilon).$$

$$\text{limit } \frac{a_2}{a^2} = K';$$

$$\frac{a_2}{a^2} = K' + \varepsilon',$$

$$a_2 = a^2(K' + \varepsilon');$$

$$a_n = a^n(K^{(n)} + \varepsilon^{(n)}).$$

Examples.

Show, by the aid of these expressions, that the limit of the ratio of any infinitesimal to one of the same order is finite; to one of a lower order is zero; to one of a higher order is infinite. That the order of the product of infinitesimals is the sum of the orders of the factors, and that the order of the quotient of infinitesimals may be obtained by subtracting the order of the denominator from the order of the numerator.

Show that, if the limit of the ratio of two infinitesimals is unity, they differ by an infinitesimal of an order higher than their own.

148. The sine of an infinitesimal angle is infinitesimal; for, as the angle approaches zero, the sine approaches zero as its limit.

If we take the angle as our principal infinitesimal, the sine is an infinitesimal of the first order; for we have seen that

$$\lim_{a \doteq 0} \left[\frac{\sin a}{a} \right] = 1, \qquad \text{(Art. 68)}.$$

The vers a is infinitesimal if a is infinitesimal, for

$$\text{vers } a = 1 - \cos a;$$

and as $a \doteq 0$, $\cos a \doteq 1$;

hence $\operatorname{vers} a \doteq 0$.

It is an infinitesimal of a higher order than the first, for we have seen that
$$\lim_{a \doteq 0} \left[\frac{1-\cos a}{a} \right] = 0, \qquad \text{(Art. 68)}.$$

Let us see if it is of the second order; that is, let us see if $\lim_{a \doteq 0} \left[\frac{1-\cos a}{a^2} \right]$ is finite. $\frac{1-\cos a}{a^2}$ assumes the form $\frac{0}{0}$ when $a = 0$, and we can find our required limit by the method of Art. 139, which gives us $\frac{1}{2}$ as the value sought. Therefore, when a is infinitesimal, $\operatorname{vers} a$ is infinitesimal of the second order.

Examples.

Taking a as the principal infinitesimal, show that
(1) $\tan a$ is an infinitesimal of the first order.
(2) $a - \sin a$ is an infinitesimal of the third order.
(3) $\tan a - a$ is of the third order.

149. Let y be any function whatever of x, *if we give* x *an infinitesimal increment* $\varDelta x$, *the corresponding increment* $\varDelta y$ *of* y *will be an infinitesimal of the same order as* $\varDelta x$, *unless for particular single values of* x.

To establish this proposition, we must show that $\lim_{\varDelta x \doteq 0} \left[\frac{\varDelta y}{\varDelta x} \right]$ is finite. $\lim_{\varDelta x \doteq 0} \left[\frac{\varDelta y}{\varDelta x} \right]$ cannot be zero, except for single values of x; for, suppose it could become and continue zero; $\lim_{\varDelta x \doteq 0} \left[\frac{\varDelta y}{\varDelta x} \right]$ is $D_x y$, and we have seen (Art. 38) that $D_x y$ shows the rate at which y is changing as x changes. If $D_x y$ becomes and remains zero, y does not change at all as x changes; and, therefore, is not a function of x, but a constant.

$\underset{\Delta x \doteq 0}{\text{limit}} \left[\dfrac{\Delta y}{\Delta x} \right]$ cannot become and continue infinite; for, in that case, $\underset{\Delta y \doteq 0}{\text{limit}} \left[\dfrac{\Delta x}{\Delta y} \right]$ would be zero, $D_y x$ would be zero, and x, regarded as a function of y, would be constant.

Since $\underset{\Delta x \doteq 0}{\text{limit}} \left[\dfrac{\Delta y}{\Delta x} \right]$ can be neither zero nor infinite, it must be finite, and Δy and Δx are of the same order.

150. *If the coördinates of the points of a curve are expressed as functions of a third variable a, the distance between two infinitely near points of the curve is an infinitesimal of the same order as the difference between the values of a to which the points correspond.*

The ordinary equations of the cycloid,

$$\left. \begin{array}{l} x = a\theta - a\sin\theta \\ y = a - a\cos\theta \end{array} \right\}$$

are a familiar example of the way in which the coördinates of points of a curve may be expressed as functions of a third variable. In the case of any curve, it is obvious that this may be done in a great variety of ways. *Any two equations containing* x, y, *and* a *that will reduce on the elimination of* a *to the ordinary equation of a given curve, can be used as equations of that curve.*

For example:

$$\left. \begin{array}{l} x = 2a \\ y = a + 2 \end{array} \right\} \text{are equivalent to } x - 2y + 4 = 0;$$

$$\left. \begin{array}{l} x = a\cos a \\ y = a\sin a \end{array} \right\} \text{are equivalent to } x^2 + y^2 = a^2;$$

$$\left. \begin{array}{l} x = a\cos a \\ y = b\sin a \end{array} \right\} \text{are equivalent to } \dfrac{x^2}{a^2} + \dfrac{y^2}{b^2} = 1;$$

$$\left.\begin{array}{l}x = a \sec a \\ y = b \tan a\end{array}\right\} \text{ are equivalent to } \frac{x^2}{a^2} - \frac{y^2}{b^2} = 1.$$

The proof of our proposition is as follows: Let a and $a + \Delta a$ be the two values of a in question, and (x,y) and $(x + \Delta x, y + \Delta y)$ be the two corresponding points. The distance D between these points will be, if we use rectangular coördinates, $\sqrt{(\Delta x)^2 + (\Delta y)^2}$. We wish to prove that $\underset{\Delta a \doteq 0}{\text{limit}} \left[\dfrac{D}{\Delta a}\right]$ is finite.

$$\frac{D}{\Delta a} = \frac{\sqrt{(\Delta x)^2 + (\Delta y)^2}}{\Delta a} = \sqrt{\left(\frac{\Delta x}{\Delta a}\right)^2 + \left(\frac{\Delta y}{\Delta a}\right)^2},$$

and, by Art. 149, $\underset{\Delta a \doteq 0}{\text{limit}} \left[\dfrac{\Delta x}{\Delta a}\right]$ and $\underset{\Delta a \doteq 0}{\text{limit}} \left[\dfrac{\Delta y}{\Delta a}\right]$ are both finite; hence $\underset{\Delta a \doteq 0}{\text{limit}} \left[\dfrac{D}{\Delta a}\right]$ is finite, and D is an infinitesimal of the same order as Δa.

151. *If two curves are so connected that the points of one correspond to the points of the other, so that when a point of the first curve is given, the corresponding point on the second is determined, the distance between two infinitely near points on the first curve is an infinitesimal of the same order as the distance between the corresponding points of the second curve.* For, if we suppose the coördinates of the points of the first curve expressed as functions of some variable a, the coördinates of the points of the second curve can also be regarded as functions of a; and, by Art. 150, each of the distances in question will be an infinitesimal of the same order as Δa, and each will therefore be of the same order as the other.

152. *If a straight line moves in a plane according to some law, so that each of its positions corresponds to some value of a variable a, the angle between two infinitely near positions of the line is an infinitesimal of the same order as the difference between the corresponding values of a.*

Suppose lines drawn through a fixed point O parallel to the moving line in its different positions. From O, with the radius unity, describe an arc. Consider any two positions of the moving line, and the corresponding lines at O, we wish to

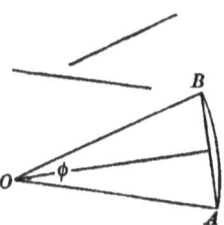

prove that the angle φ between the latter is of the same order as the difference between the values of a to which the positions of the moving line correspond. As all the lines at O correspond to values of a, the points where they cut the circle correspond to values of a, and, by Art. 150, the distance AB between two of the points supposed to be infinitely near is of the same order as $\varDelta a$. $\tfrac{1}{2}AB$ is equal to $\sin\tfrac{\varphi}{2}$; therefore $\sin\tfrac{\varphi}{2}$, and consequently $\tfrac{\varphi}{2}$ itself is an infinitesimal of the same order as $\varDelta a$, and if

$$\underset{\varDelta a \doteq 0}{\text{limit}}\left[\frac{\tfrac{\varphi}{2}}{\varDelta a}\right] \text{ is finite, } \underset{\varDelta a \doteq 0}{\text{limit}}\left[\frac{\varphi}{\varDelta a}\right] \text{ is finite.}$$

153. A simple geometrical example of an *infinitesimal of the second order* is the *perpendicular let fall upon the tangent at any point of a curve from a second point of the curve infinitely near the first.*

If, in our figure, the distance PP' is taken as the principal infinitesimal, $P'T$ is readily seen to be of a higher order than the first, for

$$\frac{P'T}{PP'} = \sin\varphi;$$

and, since $\varphi \doteq 0$ as $P' \doteq P$, its sine $\doteq 0$; hence

$$\underset{PP' \doteq 0}{\text{limit}}\left[\frac{P'T}{PP'}\right] = 0,$$

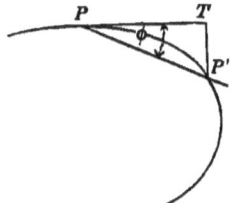

and $P'T$ is an infinitesimal of an order higher than that of PP', by Art. 147, Ex.

To show that $P'T$ is of the second order, let us consider different secant lines drawn through P, PT being itself one of these lines. Obviously, each one of these lines is determined in position when the abscissa of its second point of intersection with the curve is given; and therefore the angle between any two infinitely near secant lines, as PP' and PT is an infinitesimal of the same order as the difference between the corresponding abscissas, by Art. 152; but the distance PP' is of the same order, by Art. 150; therefore, φ and PP' are of the same order, that is, of the first order; $\sin \varphi$ is also of the first order, by Art. 148; hence $P'T$, which is equal to $PP' \sin \varphi$, is of the second order (Art. 147, Ex.).

154. *To determine the tangent at any given point of a curve*, we draw a secant line through the point in question and any second point on the curve, and seek the limiting position approached by this line as the second point approaches the first; or, in other words, *we seek the limiting position of the line joining the given point with an infinitely near point of the curve.* It can be shown that this is also the *limiting position of any line passing through the given point and a point whose distance from the second point of the curve is an infinitesimal of a higher order than the distance between the two points on the curve.*

Let P and P' be two infinitely near points on the curve, and let $P'M$ be an infinitesimal of a higher order than PP', then the limiting position of PP' as $P' \doteq P$ will be the same as the limiting position of PM; for, in the triangle PMP',

$$\frac{P'M}{PP'} = \frac{\sin\varphi}{\sin\psi};$$

hence
$$\sin\varphi = \frac{P'M}{PP'}\sin\psi;$$

and as by hypothesis, $\displaystyle\lim_{PP' \doteq 0}\left[\frac{P'M}{PP'}\right] = 0,$

$\displaystyle\lim_{PP' \doteq 0}[\sin\varphi]$ must be zero. Therefore

$$\lim_{PP' \doteq 0}[\varphi] = 0,$$

and the two lines, PP' and PM, approach the same limiting position.

155. This principle is frequently of service in problems concerning the position of tangent lines. For example: *Suppose perpendiculars let fall from a fixed point to the tangents of a given curve, to draw the tangent at any given point of the locus on which the feet of these perpendiculars lie.*

Let M and M' be two infinitely near points of the given curve, and O be the given point from which the perpendiculars are let fall; then P and P' are two infinitely near points of the locus in question, and the required tangent at P is the limiting position of the line joining P and P'. Draw through M the line MP''

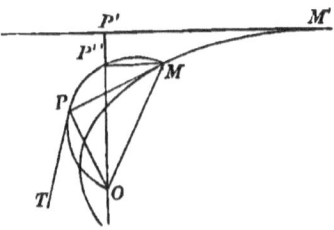

parallel to the tangent $M'P'$. If we take MM' as our principal infinitesimal, $P''P'$ is an infinitesimal of the second order, by Art. 153, and PP' is of the first order, by Art. 151; consequently (Art. 154) it will answer our purpose to find the limiting position of the line joining PP''; but, since $MP''O$ and MPO are both right angles, P'' lies on the circumference of a circle described on OM as diameter, and the required limiting position of PP'' is that of a tangent to this circle at P, which is therefore the required tangent. Hence to obtain a tangent to the locus in question at any given point, we have only to join the corresponding point with O, to erect a circle on this joining line as diameter, and to draw a tangent to the circle at the given point. Of course, the normal to this locus at the given point bisects the joining line OM.

156. Let us consider the locus of the feet of perpendiculars let fall from the focus of an ellipse upon the tangents to the curve.

Since the tangent to the required locus at P is tangent to the circle on FM as diameter, the normal at P passes through the

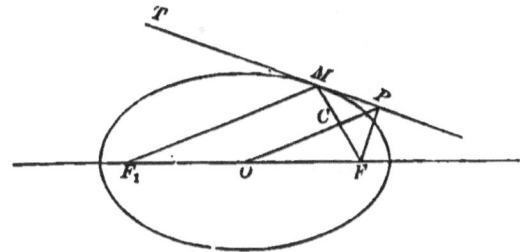

centre C of the circle. Draw the focal radius $F_1 M$. Since the tangent to an ellipse makes equal angles with the focal radii drawn to the point of contact,

$$TMF_1 = PMC;$$

$$PMC = MPC,$$

because MC and CP are equal;

$$\therefore MPC = TMF_1,$$

and PO is parallel to MF_1; it must then divide MF and $F_1 F$ proportionally; and as it bisects MF, it also bisects $F_1 F$, and consequently passes through the centre of the ellipse. Since every normal to the required locus passes through the centre of the ellipse, the locus is a circle concentric with the ellipse. It is easily seen that it must pass through the vertices of the ellipse. It is then a circle on the major axis of the ellipse as diameter.

Example.

Show that the locus of the foot of a perpendicular let fall from the focus upon any tangent is a circle on the transverse axis as diameter in the hyperbola; is the tangent at the vertex in the parabola.

Problem.

157. *Upon each normal to a plane curve a point is taken at a constant distance from the intersection of the normal with the curve;* to find *the tangent at any point of the locus thus formed.*

Let M and M' be two infinitely near points on the given curve, P and P' the corresponding points of the locus; let

$$MP = M'P' = a;$$

call the angle between the normals, φ. Draw MM'' and PP'' perpendicular to the second normal. The required tangent is the

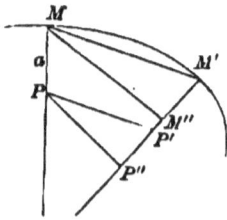

limiting position of PP', and the tangent at M is the limiting position of MM'. If MM' is taken as the principal infinitesimal, PP' and φ are of the first order and $M'M''$ of the second (Arts. 151–153). $P'P''$ is of an order higher than the first, for

$$P'P'' = M'M'' + M''P'' - a,$$
$$M''P'' = a\cos\varphi;$$
hence $\quad P'P'' = M'M'' - a(1-\cos\varphi).$

φ being of the first order, $1 - \cos\varphi$ is of the second order by Art. 148; and as $M'M''$ is of the second order, $P'P''$ is of at least as high an order as the second. By Art. 154, our required tangent will be the limiting position of PP'', and the tangent at M will be the limiting position of MM''; but PP'' and MM'' are parallel always; therefore their limiting positions are parallel, and our required tangent is parallel to the tangent to the given curve at the corresponding point, and the curves are what are called parallel curves.

Problem.

158. *An angle of constant magnitude is circumscribed about a given curve; to draw a tangent to the locus of its vertex.*

The required tangent is the limiting position of the secant line PP'. Draw through M and N lines MP'', NP'', parallel to the tangents at M and N. It can be shown that the sides, and therefore the diagonal, of the parallelogram $P'P''$ are infinitesimals of a higher order than PP', and therefore that the re-

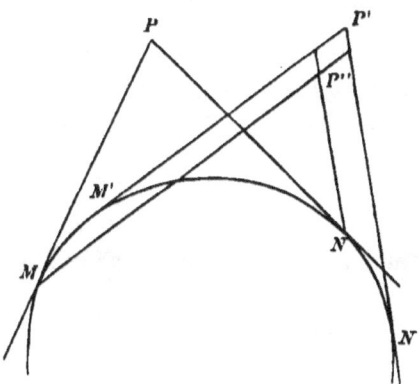

quired tangent can be found as the limiting position of PP'''. Since the angles at P and P''' are equal, the point P''' lies on a circle circumscribed about MPN; the limiting position of PP''' is therefore the tangent to this circle at P. Our solution is, then, draw a circle through the vertex of the circumscribing angle and the points of contact of its sides, and the tangent to this circle at the vertex of the angle is the tangent required.

EXAMPLE.

Show that the locus of the vertex of a right angle circumscribed about an ellipse or an hyperbola is a concentric circle; about a parabola is the directrix.

159. In the preceding examples, the advantage we have gained in the use of infinitesimals has arisen from the fact that we have been able to replace one infinitesimal by another related to it and more simply connected with the other values considered in the problem. The possibility of such substitutions, and the limitations under which they can be made, form the subject of the following two theorems, which are of prime importance, and lie at the foundation of the *Infinitesimal Calculus*.

THEOREM.

160. *In any problem concerning the limit of the ratio of two infinitesimals, either may be replaced by any infinitesimal so related to it that the limit of the ratio of the second to the first is unity.*

PROOF.

Let a, β, a', and β' be infinitesimals so related that

$$\operatorname{limit} \frac{a'}{a} = 1 \text{ and } \operatorname{limit} \frac{\beta'}{\beta} = 1.$$

Then will

$$\operatorname{limit} \frac{a}{\beta} = \operatorname{limit} \frac{a'}{\beta'}.$$

$$\frac{a}{\beta} = \frac{a'}{\beta'} \cdot \frac{a}{a'} \cdot \frac{\beta'}{\beta} \qquad \text{identically};$$

hence
$$\operatorname{limit}\frac{a}{\beta} = \operatorname{limit}\frac{a}{\beta'} \times \operatorname{limit}\frac{a}{a'} \times \operatorname{limit}\frac{\beta'}{\beta},$$

$$\operatorname{limit}\frac{a}{\beta} = \operatorname{limit}\frac{a'}{\beta'} \times 1 \times 1 = \operatorname{limit}\frac{a'}{\beta'}.$$
Q.E.D.

THEOREM.

161. *In any problem concerning the limit of a sum of infinitesimals, provided that this limit is finite, any infinitesimal may be replaced by another so related to it that the limit of the ratio of the second to the first is unity.*

PROOF.

Let
$$a_1 + a_2 + a_3 + \cdots + a_n$$
be a sum of infinitesimals of such a nature that the number of the terms increases as each term decreases in absolute value, so that the limit of the sum is some finite quantity.

Let $\beta_1, \beta_2, \beta_3, \cdots \beta_n$ be a set of infinitesimals so related to the first set that
$$\operatorname{limit}\frac{\beta_1}{a_1} = 1, \ \operatorname{limit}\frac{\beta_2}{a_2} = 1, \ \&c., \ \operatorname{limit}\frac{\beta_n}{a_n} = 1,$$

then
$$\frac{\beta_1}{a_1} = 1 + \epsilon_1, \ \frac{\beta_2}{a_2} = 1 + \epsilon_2, \ \&c., \ \frac{\beta_n}{a_n} = 1 + \epsilon_n,$$

$\epsilon_1, \epsilon_2, \cdots \epsilon_n$ being necessarily infinitesimal (Art. 7).

$$\beta_1 = a_1 + a_1 \epsilon_1,$$
$$\beta_2 = a_2 + a_2 \epsilon_2,$$
$$\beta_n = a_n + a_n \epsilon_n,$$
$$\beta_1 + \beta_2 + \beta_3 + \cdots + \beta_n = a_1 + a_2 + a_3 + \cdots + a_n$$
$$+ a_1 \epsilon_1 + a_2 \epsilon_2 + a_3 \epsilon_3 + \cdots + a_n \epsilon_n.$$

Let η be such a variable that at any instant it shall be equal to the greatest in absolute value of the quantities $\epsilon_1, \epsilon_2, \ldots \epsilon_n$. Of course, since each of these approaches zero as its limit, η must also approach zero as its limit; i.e., η is infinitesimal.

$$a_1\epsilon_1 + a_2\epsilon_2 + a_3\epsilon_3 + \cdots + a_n\epsilon_n < \eta(a_1 + a_2 + a_3 + \cdots + a_n),$$

hence $\beta_1 + \beta_2 + \beta_3 + \cdots + \beta_n - (a_1 + a_2 + a_3 + \cdots + a_n)$

$$< \eta(a_1 + a_2 + a_3 + \cdots + a_n).$$

By hypothesis, $\quad \text{limit}\,(a_1 + a_2 + \cdots + a_n) \quad$ is finite;

therefore, limit of $\;\eta(a_1 + a_2 + a_3 + \cdots + a_n) \quad$ is zero.

Consequently

$$\text{limit}\,(\beta_1 + \beta_2 + \beta_3 + \cdots + \beta_n) = \text{limit}\,(a_1 + a_2 + a_3 + \cdots + a_n).$$

Q.E.D.

162. *If two infinitesimals differ from each other by an infinitesimal of a higher order, the limit of their ratio is unity.*

For, let $\quad a' - a = \epsilon,$

where ϵ is of a higher order than a;

$$a' = a + \epsilon,$$

$$\frac{a'}{a} = 1 + \frac{\epsilon}{a},$$

$$\text{limit}\,\frac{a'}{a} = 1 + \text{limit}\,\frac{\epsilon}{a};$$

but, by hypothesis, $\quad \text{limit}\,\dfrac{\epsilon}{a} = 0, \quad$ (Art. 147, Ex.);

therefore $\quad \text{limit}\,\dfrac{a'}{a} = 1.$

It follows that the theorems of Art. 160 and Art. 161 can be stated as follows:—

In finding the limit of a ratio, or the limit of a sum of infinitesimals, any infinitesimal may be replaced by one that differs from it by an infinitesimal of a higher order. Or, in finding the limit of a ratio or of a sum of infinitesimals, any infinitesimal term may be neglected without in the least affecting the result, provided that it is of a higher order than the terms retained.

163. Let us take the problem of *finding the direction of the tangent to a parabola.*

The tangent $T'T$ at P is the limiting position of the secant through P and P'. Draw the focal radii FP and FP', and the perpendiculars PR and $P'S$ to the directrix. Draw PM and

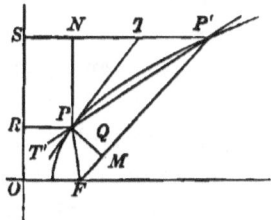

PN perpendicular to FP' and $P'S$ respectively, and with F as a centre, and with the radius FP, describe the arc PQ.

Take PP' as the principal infinitesimal, then $P'M$ and $P'N$ are of the first order, since the limit of the ratio of each of them to PP' is finite.

PQ is of the first order, by Art. 151, and MQ is of the second order, by Art. 153.
$$P'S = P'F,$$
from the definition of a parabola;
$$PR = PF = QF;$$
$$\therefore P'N = P'Q.$$
$$\cos PP'F = \frac{P'M}{PP'},$$

$$\cos PP'S = \frac{P'N}{PP'},$$

$$\frac{\cos T'PF}{\cos T'PR} = \underset{P' \doteq P}{\text{limit}} \left[\frac{\cos PP'F}{\cos PP'S} \right] = \underset{P' \doteq P}{\text{limit}} \left[\frac{P'M}{P'N} \right]$$

$$= \underset{P' \doteq P}{\text{limit}} \left[\frac{P'Q}{P'N} \right] = 1, \qquad \text{by Art. 162;}$$

$$\therefore T'PR = T'PF,$$

and the tangent at any point of a parabola bisects the angle between the focal radius and the diameter through the given point.

164. *To find the area of the sector of a parabola included between two focal radii.* Take points of the parabola between the extremities of the bounding radii, and join them with the focus, thus dividing the area in question into smaller sectors, of which the sector FPP' in the figure of the last article may be taken as a type. Draw perpendiculars from the extremities of the bounding radii to the directrix, and consider the external area bounded by them, the directrix and the curve; draw perpendiculars from the intermediate points already described to the directrix, and the external area will be divided into smaller curvilinear quadrilaterals, of which $PP'RS$ is one. No matter how close together the intermediate points are taken, the external area is the actual sum of these small curvilinear quadrilaterals; it is then the limit of their sum as the number is indefinitely increased. If the distance between any two of the points, as PP', is taken as the principal infinitesimal, PN, $P'N$, PM, $P'M$, are all infinitesimals of the first order, since the limit of the ratio of each of them to PP' is the sine or the cosine of a finite angle. The area of $PP'RS$ lies between $P'S \times PN$ and $NS \times PN$, and is therefore an infinitesimal of the first order. Hence we have to consider the limit of a sum of infinitesimals where the limit is finite, and we can replace any one by one differing from it by an infinitesimal of a higher order than the first. The rectangle $PRSN$ differs from

$PP'RS$ by less than a rectangle on PN and $P'N$; that is, by less than $PN \times P'N$, an infinitesimal of the second order. Therefore the required external area, which is the limit of the sum of infinitesimal areas of which $PP'RS$ is a type, and which we shall indicate by limit $\Sigma PP'RS$ (Σ serving as a symbol for the word *sum*), is equal to limit $\Sigma PRNS$.

The given sector is equal to the sum of the smaller sectors of which FPP' is a type = limit $\Sigma FPP'$, each term here being an infinitesimal of the first order. Draw the straight line PQ. The triangle FPQ differs from the sector FPP' by less than a rectangle on PM and MP', which would be of the second order, and may therefore replace FPP' in the expression for our required area.

$$\text{limit}\frac{PN}{PM} = 1, \qquad \text{by Art. 163};$$

consequently PM and PN differ by an infinitesimal of higher order than the first, and the triangle FPQ differs from one-half the rectangle $PRNS$ by an infinitesimal of higher order than the first, and may be replaced by $\frac{1}{2}PRNS$.

We have then, external area = limit $\Sigma PRNS$,

given sector = limit $\Sigma \frac{1}{2} PRNS$;

and the given focal sector is equal to one-half the area bounded by the curve, the directrix and perpendiculars let fall from the extremities of the arc of the given sector to the directrix.

Infinitesimal Arc and Chord.

165. Let us consider the relation between the lengths of an infinitesimal chord and its arc.

Take the chord PP' as the principal infinitesimal, and draw the tangents PT and $P'T$. The arc PP' is less than $PT + P'T$ and greater than the chord PP'. The angles ϵ and ϵ' are infinitesimal.

$$\cos \varepsilon = \frac{PM}{PT},$$

$$\cos \varepsilon' = \frac{P'M}{P'T},$$

$$\text{limit } \cos \varepsilon = 1,$$

and $\quad\quad\quad$ limit $\cos \varepsilon' = 1$;

therefore $\quad\quad$ limit $\dfrac{PM}{PT} = 1$

and $\quad\quad\quad$ limit $\dfrac{P'M}{P'T} = 1.$

Thus $\quad\quad\quad PM = PT + \eta$

and $\quad\quad\quad P'M = P'T + \eta',$

where η and η' are infinitesimals of a higher order than the first, by Art. 147, Ex.

$$PM + P'M = PT + P'T + \eta + \eta',$$

or the difference between the sum of the tangents and the chord is of a higher order than the first. The difference between the arc and the chord is less than this, therefore *the limit of the ratio of an infinitesimal arc to its chord is unity.*

166. *It is customary to say roughly that lines which make with each other an infinitesimal angle*, that is, *lines which approach the same limiting position, coincide, and that finite values which differ by an infinitesimal or infinitesimal values which differ by an infinitesimal of a higher order*, that is, *values such that the limit of their ratio is unity, are equal;* and this way of speaking is very convenient, especially for preliminary investigations. It is important, however, to be able to put a proof given in this form into the more exact language of limits.

It is easily seen from what has just been said, that *the line* joining two infinitely near points of any curve, can, speaking roughly, be regarded at pleasure as chord, arc, or tangent, so that an infinitesimal arc can be treated as a straight line.

167. As an example of this loose form of proof, let us show that a tangent to an ellipse makes equal angles with the focal radii drawn to the point of contact.

Let P and P' be two infinitely near points of the ellipse, then PP' is the tangent in question. From F and F' as centres, draw

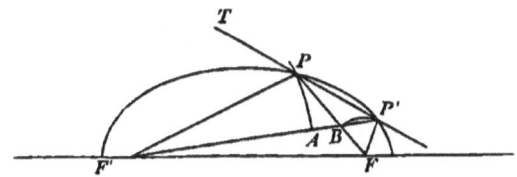

the arcs PA and $P'B$; PA and $P'B$ being infinitesimal arcs, are straight lines, and PAP' and $P'BP$ are right angles, since the tangent to a circle is perpendicular to the radius drawn to the point of contact.
$$F'P + PF = F'P' + P'F,$$
by the definition of an ellipse. Take away from the first sum $F'P + BF$, and we have left PB; take away from the second sum the equal amount $F'A + P'F$, and we have left $P'A$;
$$\therefore PB = P'A;$$
and the right triangles PAP' and PBP' have the hypothenuse and a side of the one equal to the hypothenuse and a side of the other, and are equal; and the angle
$$FPP' = F'P'P;$$
but the lines $F'P'$ and $F'P$ coincide, so that the angle $F'P'P$ is the same as the angle $F'PT$; and
$$\therefore F'PT = FPP',$$
and the tangent makes equal angles with the focal radii. Q.E.D.

EXAMPLE.

Prove that a tangent to an hyperbola bisects the angle between the focal radii drawn to the point of contact.

168. To find the area of a segment of a parabola cut off by a line perpendicular to the axis. Compare the required area with the area of the circumscribing rectangle. We can regard the

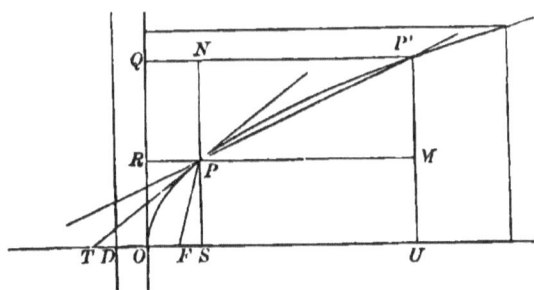

first as made up of the infinitesimal rectangles of which $PMUS$ is a type, and the second of the corresponding rectangles of which $QNPR$ is one. Draw the directrix.

$$PF = SD \text{ and } DO = OF,$$

by the definition of the parabola; but

$$PF = FT \qquad \text{by Art. 163};$$

$$\therefore TO = OS.$$

The triangles $P'MP$ and PST are similar, and

$$\frac{P'M}{PS} = \frac{PM}{ST} = \frac{PM}{2OS};$$

hence $\qquad PM \times PS = 2OS \times P'M = 2PRQN,$

or rectangle $\qquad PU = 2PQ;$

$$\therefore \Sigma PU = 2\Sigma PQ,$$

and the segment in question is twice the external portion of the circumscribing rectangle, and, therefore, is two-thirds of the whole rectangle.

EXAMPLE.

Prove the theorems of Arts. 167, 168, strictly, by the method of limits.

169. The properties of the cycloid can be very simply and neatly obtained by the aid of infinitesimals; though, for this purpose, it is better to look at the curve from a new point of view.

Let a fixed circle equal to the generating circle be drawn tangent to the base of the cycloid at its middle point; through the

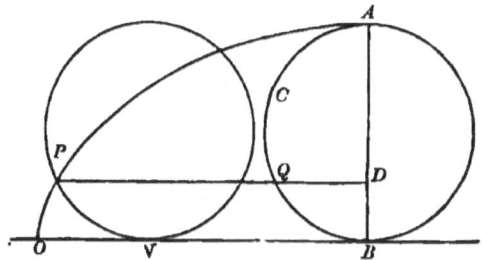

generating point P, draw PQD parallel to the base. From the nature of the cycloid, the arc

$$PN = ON \text{ and } OB = ACB,$$

$$PQ = NB = OB - ON = ACB - QB = ACQ.$$

Hence *points of the cycloid can be obtained by erecting perpendiculars to a diameter of a fixed circle, and extending each until its external portion is equal to the distance along the arc of the circle from the perpendicular in question to a given end of the diameter.*

170. *The tangent to the cycloid passes through the highest point of the generating circle.*

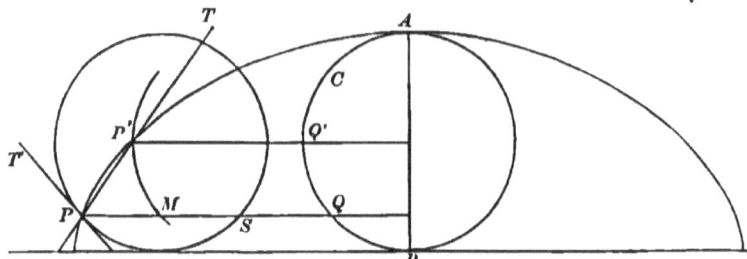

Rough Proof. — Let P and P' be infinitely near, then PP' is the required tangent; through P' draw an arc parallel and similar to QQ'. This arc may be regarded as a straight line. The triangle $PP'M$ is isosceles, since

$$QP = ACQ \text{ and } MQ = P'Q' = ACQ',$$

hence $\qquad PM = QQ' = MP';$

$\therefore$ the angle $PP'M = P'PM.$

$P'M$ is parallel to the tangent at P to the generating circle, hence

$$PP'M = TPT',$$

and PT bisects the angle MPT', bisects the arc PTS, and consequently passes through the highest point of the generating circle. Q.E.D.

Strict Proof. — Draw the chord $P'M$, and regard PP' as a secant line; in the triangle $PP'M$ we have

$$\frac{\sin PP'M}{\sin TPM} = \frac{PM}{P'M},$$

$$\therefore \text{ limit } \frac{\sin PP'M}{\sin TPM} = \text{limit } \frac{PM}{P'M}.$$

The $\qquad$ arc $P'M = PM,$

and the chord $P'M$ differs from the arc by an infinitesimal of a higher order than that of the chord.

$$\therefore \text{ limit } \frac{PM}{\text{chord } P'M} = \text{limit } \frac{PM}{\text{arc } P'M} = 1,$$

hence $\qquad\qquad \text{limit } PP'M = \text{limit } TPM.$

The limiting position of $P'M$ is the tangent PT';

$$\therefore \text{ limit } TPQ = \text{limit } TPT',$$

and the tangent passes through the highest point of the generating circle.

The Area of the Cycloid.

171. *Rough Investigation.* — Circumscribe a rectangle about the cycloid, and its area is evidently equal to the circumference of the generating circle multiplied by its diameter; that is, to four times the area of the circle. The area of the cycloid is

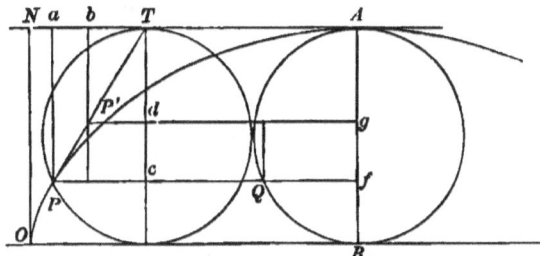

this area minus the area of the external portion of the rectangle. The external area ANO may be divided into trapezoids, of which $abPP'$ is any one. The tangent PP' passes through the highest point of the generating circle, and is a diagonal of the rectangle $TaPc$, Tc being a diameter. From geometry,

$$abP'P = cdP'P,$$

which is equal to Qg; therefore the sum of the trapezoids $abP'P$ is equal to the sum of the corresponding rectangles Qg, or the external area ANO is equal to the semi-circle ACB: but ANO

is half of the external portion of the circumscribing rectangle; consequently, the *area of the cycloid is three times the area of the generating circle.*

Strict Proof. — The external area is the sum of the curvilinear quadilaterals of which $abP'P$ is any one; that is,

$$\text{area} = \Sigma abP'P = \text{limit } \Sigma abP'P = \text{limit } \Sigma abhP,$$

for $$abhP - abP'P < eP'hP,$$

which is of the second order. $P'P''$ is of the second order, since

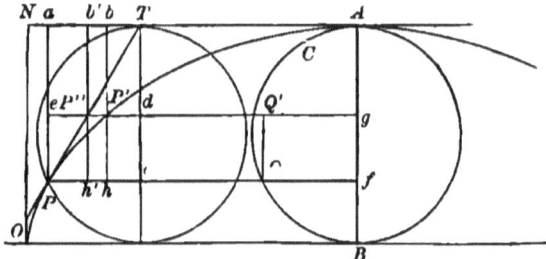

it is proportional to the distance from P' to the tangent at P (Art. 153); therefore $bhh'b'$ is of the second order, and

$$\text{limit } \Sigma abhP = \text{limit } \Sigma ab'h'P.$$

$$ab'h'P = edcP = Qg,$$

hence the external area $= \text{limit } \Sigma Qg = $ area of ACB.

Length of an Arc of the Cycloid.

172. *Rough Proof.* — The arc AP is equal to the sum of the infinitesimal chords of which PP' is one. The chord AQ is the sum of the differences between each chord and the one drawn to a point of the fixed circle above the point in question and infinitely near it; QS is such a difference, hence

$$\text{arc } AP = \Sigma PP' \text{ and chord } AQ = \Sigma QS.$$

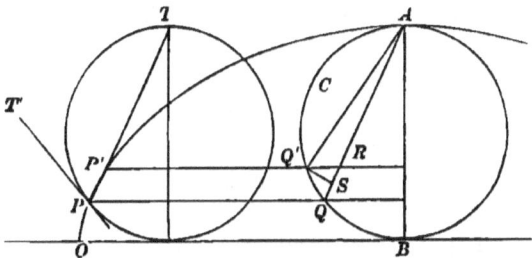

PP' and QR are equal, $Q'QR = Q'RQ$, by Art. 170, and $QQ'R$ is isosceles. $Q'S$, an infinitesimal arc described from A as a centre, may be regarded as a straight line perpendicular to QR, and therefore bisects QR, and

$$PP' = 2QS,$$

$$\Sigma PP' = 2\Sigma QS.$$

$$\text{Arc } AP = 2 \text{ chord } \dot{A}Q.$$

The $\text{arc } AO = 2AB$,

and *the whole arc of the cycloid is eight times the radius of the generating circle.*

Strict Proof.—$P'P''$, $Q'T'''$, and US are infinitesimals of the second order, each being proportional to the distance from a point

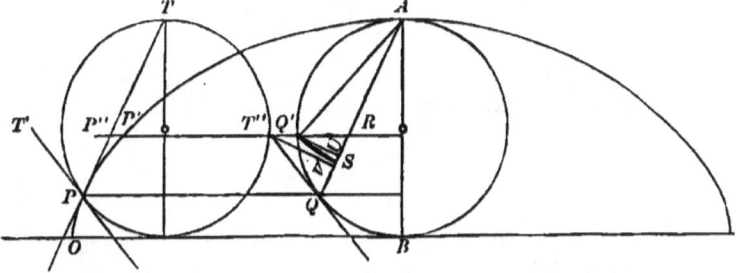

of a curve to the tangent at a point infinitely near. Vv is also of second order, as it is the projection of $Q'T'''$ on AQ.

The arc AP = limit $\Sigma PP'$ = limit $\Sigma PP''$,

since, in triangle $PP'P''$,

$$PP' - PP'' < P'P''.$$

The chord $AQ = \Sigma QS$;

$$= \text{limit } \Sigma QS = \text{limit } \Sigma QU = \text{limit } \Sigma QV.$$

But the triangle $QT''R$ is isosceles, hence

$$QV = \tfrac{1}{2} QR = \tfrac{1}{2} PP'';$$

and, as arc AP = limit $\Sigma PP''$,

$$\text{arc } AP = 2 \text{ chord } AQ. \qquad \text{Q.E.D.}$$

Radius of Curvature of the Cycloid.

173. Rough Investigation. — The centre of curvature for P is the intersection of the normal at P with the normal at P'.

PX, $P'X$, and PP' are parallel to QB, $Q'B$, and QS respectively, hence the triangles $PP'X$ and QSB are similar. The angle

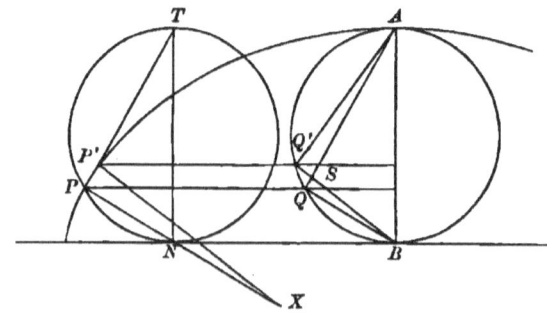

Q is a right angle, the angle B is infinitesimal; the angle QSB differs from a right angle by an infinitesimal, and may be regarded as a right angle. Therefore, by Art. 172,

$$QS = \tfrac{1}{2} PP',$$

and consequently, $QB = \tfrac{1}{2} PX$;

and the radius of curvature is twice PN, *the portion of the normal within the generating circle.*

Strict Proof. — The centre of curvature is the limiting position of X.

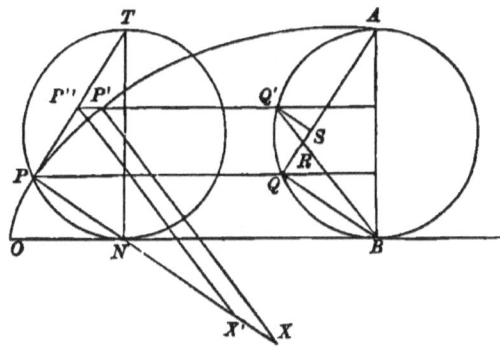

$PP''X'$ is similar to QRB, hence

$$\frac{PX'}{QB} = \frac{PP''}{QR} \text{ and limit } \frac{PX'}{QB} = \text{limit } \frac{PP''}{QR}. \qquad (1)$$

Let PP' be the principal infinitesimal, then $P'P''$ is of the second order; therefore, in (1), PX can be substituted for PX'. $RQ'S$ is similar to BQR, hence $\dfrac{RS}{QR} = \dfrac{Q'S}{QB}$.

QR and QS are infinitesimal, QB is finite, RS is of the second order, and QS can be substituted for QR in (1), and

$$\text{limit } \frac{PX}{QB} = \text{limit } \frac{PP''}{QS} :$$

but, by Art. 172, $\text{limit } \dfrac{PP''}{QS} = 2$;

$$\therefore \text{ limit } PX = 2QB = 2PN. \qquad \text{Q.E.D.}$$

Evolute of the Cycloid.

174. Extend the diameter TN to N', making

$$NN' = TN$$

and draw a circle on NN' as diameter. The centre of curvature X, corresponding to P, will lie on this circle, since

$$PX = 2PN.$$

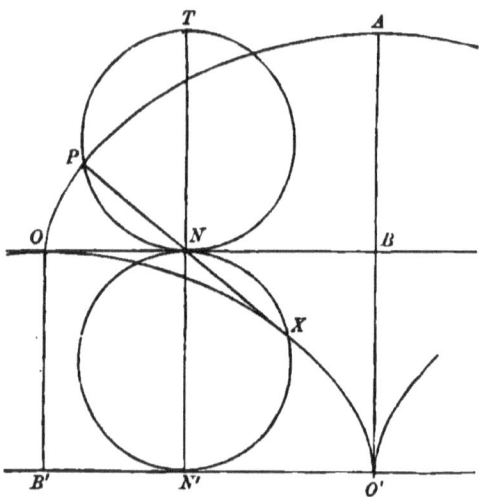

Draw a tangent to the second circle at N', drop a perpendicular from O to this tangent, and lay off $B'O'$ equal to one-half the circumference of the generating circle.

The $\qquad \text{arc } PN = ON = B'N';$

$$\therefore \text{ the arc } N'X = N'O',$$

and X lies on a cycloid equal to the given cycloid, having its origin at O' and its highest point at O, and this must be the evolute required.

EXAMPLES.

175. (1) From a point O situated in the plane of a plane curve, radii vectores are drawn to different points of the curve, and on each one a distance is laid off from O inversely proportional to the length of the radius vector; to determine the tangent at any point of the locus of the points thus obtained.

(2) Take any two curves in the same plane, and consider as corresponding points those at which the tangents are parallel; draw through a fixed point lines equal and parallel to those uniting corresponding points of the two curves. Prove that a tangent to the locus of the points thus obtained is parallel to the tangents at the corresponding points of the given curves, and that any arc of this curve is the sum or difference of those which correspond to it upon the given curves.

(3) From a point O radii vectores are drawn to a given curve, and each is extended beyond the curve by a constant length. Prove that the normal to the curve on which the extremities of the radii vectores lie, the normal at the corresponding point of the given curve, and the perpendicular through O to the radius vector of the point, have common intersection.

176. To show the power of this method of infinitesimals, we shall give an investigation into the nature of what is called the *Brachistochrone*, or *Curve of Quickest Descent*. The problem is a famous one, and the solution below is in effect the one given by James Bernouilli, and is very much simpler and more elementary than the usual analytical solution which requires the use of the *Calculus of Variations*.

The problem is, *given two points not in the same horizontal plane, nor in the same vertical line; to find the curve down which a particle moving without friction can slide in the least time from the upper point to the lower, the accelerating force being terrestrial gravitation.*

Let us first consider a simpler question: To find the path of quickest descent on the hypothesis that it is to consist of *two*

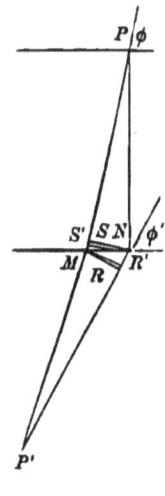

straight lines intersecting on a given horizontal plane, assuming that the particle moves down each line with a uniform velocity equal to the mean velocity with which it would actually descend the line in question. It is easily seen that both lines must lie in the vertical plane containing the two given points.

Let PNP' and PMP' be two paths of *equal* time from P to P'. Then the required path must lie between them. If we suppose them to approach, continuing still paths of equal time, the required path of quickest descent will be the limiting position of either of them. Let v be the mean velocity of a particle sliding from P to M; then, by Art. 115, v will also be the mean velocity of a particle sliding from P to N.

Let v_1 be the mean velocity of a particle sliding from M to P', supposing that the particle started from M with the velocity actually acquired by sliding down PM; then v_1 is also the mean velocity of descent from N to P', by Art. 115. As we are going to make the paths PMP' and PNP' approach indefinitely, MN is an infinitesimal. Draw the arcs NS' and MR' from P and P' as centres, and the perpendiculars NS and MR. On our hypothesis, the time of descent from P to S' equals time of descent from P to N, and time of descent from M to P' equals time of descent from R' to P'; hence, as time PMP' equals time PNP', the time of descent from S' to M equals time from N to R',

or
$$\frac{S'M}{v} = \frac{NR'}{v_1},$$

whence
$$\frac{S'M}{NR'} = \frac{v}{v_1},$$

and
$$\operatorname{limit} \frac{S'M}{NR'} = \operatorname{limit} \frac{v}{v_1};$$

$$\operatorname{limit} \frac{S'M}{NR'} = \operatorname{limit} \frac{SM}{NR},$$

CHAP. X.] INFINITESIMALS. 179

$$\frac{SM}{NR} = \frac{\cos PMN}{\cos P'MN} = \frac{\cos \varphi}{\cos \varphi_1};$$

hence
$$\text{limit}\frac{\cos \varphi}{\cos \varphi_1} = \text{limit}\frac{v}{v_1}.$$

Let the angles made with the horizontal by the two portions of the required path be θ and θ_1, and the mean velocities down the two portions of the required path be v and v_1. Then

$$\frac{\cos \theta}{\cos \theta_1} = \frac{v}{v_1} \text{ or } \frac{\cos \theta}{v} = \frac{\cos \theta_1}{v_1}.$$

Let us now consider a path of quickest descent, consisting of *three rectilinear portions intersecting on given horizontal planes*, all the other conditions remaining as before. Let $PRSP'$ be the required path. It is easily seen that PRS must be the path of quickest descent under the given conditions from P to S; so that

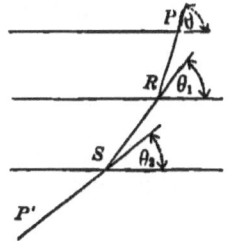

$$\frac{\cos \theta}{v} = \frac{\cos \theta_1}{v_1}.$$

RSP' must be the path of quickest descent from R to P' under the given conditions, so that

$$\frac{\cos \theta_1}{v_1} = \frac{\cos \theta_2}{v_2},$$

v, v_1, v_2 being mean velocities down PR, RS, and SP', respectively.

Suppose now that the number of rectilinear portions of the broken line of descent is indefinitely increased, each portion will decrease indefinitely in length, and the path will approach a curve as its limiting form. The mean velocity down each portion of the polygonal path will approach as its limit the actual velocity at the corresponding point of the limiting curve; the angle made by each portion with the horizontal will approach the angle made by the curve with the horizontal: hence our limiting curve, which is obviously the *required brachistochrone*, must be of such a nature that *the cosine of the angle it makes at each point with the horizontal shall be proportional to the velocity the particle will possess on reaching that point.* Let us take the

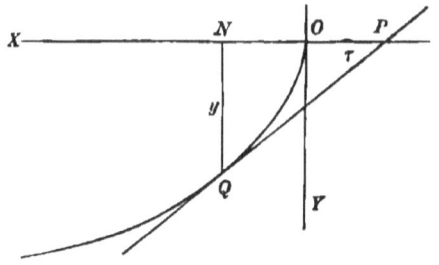

horizontal and vertical lines through the highest given point as our axes, and take the positive directions of X and Y as the usual negative directions. The velocity acquired by a particle sliding from O to Q is, by Art. 118, the velocity it would acquire falling from N to Q, that is, $\sqrt{(2gy)}$. We shall have then, as the defining property of the required curve,

$$\frac{\cos \tau}{\sqrt{(2gy)}} = K,$$

where K is a constant; or $\cos \tau = Cy^{\frac{1}{2}}$,

C being some constant. *The cycloid is a curve possessing this property,* as is easily seen.

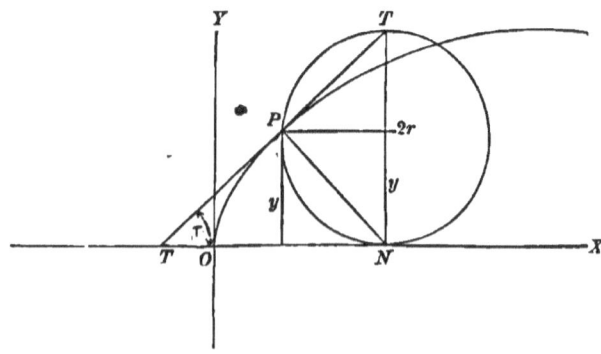

INFINITESIMALS.

We have $$\cos\tau = \sin PT'N = \frac{PN}{2r};$$

but, by geometry, $$PN = \sqrt{(2ry)};$$

hence $$\cos\tau = \frac{\sqrt{(2ry)}}{2r} = \sqrt{\left(\frac{y}{2r}\right)} = \frac{1}{\sqrt{(2r)}} y^{\frac{1}{2}}. \quad \text{Q.E.D.}$$

The converse, that every curve possessing the property

$$\cos\tau = Cy^{\frac{1}{2}}$$

is a *cycloid*, can be proved analytically by finding its equation, as follows:—

Let the required equation be

$$y = fx.$$

We know that $$\tan\tau = D_x y,$$

$$\cos\tau = \frac{1}{\sqrt{1+(D_x y)^2}} = Cy^{\frac{1}{2}};$$

$$1 = C^2 y [1 + (D_x y)^2],$$

$$(D_x y)^2 = \frac{1 - C^2 y}{C^2 y}.$$

Call $$C = \frac{1}{\sqrt{(2a)}}$$

and assume $$y = a - a\cos\theta.$$

$$D_x y = \frac{D_\theta y}{D_\theta x},$$

and we have

$$\frac{a^2 \sin^2\theta}{(D_\theta x)^2} = \frac{1 - \dfrac{1-\cos\theta}{2}}{\dfrac{1-\cos\theta}{2}} = \frac{1+\cos\theta}{1-\cos\theta}.$$

$$(D_\theta x)^2 = \frac{a^2 \sin^2\theta (1-\cos\theta)}{1+\cos\theta} = \frac{a^2(1-\cos^2\theta)(1-\cos\theta)}{1+\cos\theta}$$

$$= a^2(1-\cos\theta)^2,$$

$$D_\theta x = a(1-\cos\theta),$$

$$x = a\smallint_\theta (1-\cos\theta) = a\theta - a\sin\theta + C,$$

when $\qquad x=0, \; y=0, \text{ and } \theta = 0;$

hence $\qquad\qquad C=0,$

and our equations are $\quad \left. \begin{aligned} x &= a\theta - a\sin\theta \\ y &= a - a\cos\theta \end{aligned} \right\}$

the familiar equations of a cycloid; and the *brachistochrone is an inverted cycloid with its cusp at the higher of the given points.*

CHAPTER XI.

DIFFERENTIALS.

177. A DERIVATIVE has, in effect, been defined as *the limit of the ratio of infinitesimal increments of function and variable*. Consequently, in getting a derivative, we can replace the increment of the function by any quantity differing from it by an infinitesimal of a higher order.

For example: in getting $D_x x^2$, we find

$$\Delta(x)^2 = x^2 + 2x\Delta x + (\Delta x)^2 - x^2 = 2x\Delta x + (\Delta x)^2.$$

$2x\Delta x$ differs from $\Delta(x^2)$ by $(\Delta x)^2$, which is of the second order if we take Δx as the principal infinitesimal, and $2x\Delta x$ may be substituted for $\Delta(x^2)$ in getting $D_x x^2$, which then equals

$$\lim_{\Delta x \doteq 0} \left[\frac{2x\Delta x}{\Delta x} \right] = \lim_{\Delta x \doteq 0} [2x] = 2x.$$

In our old problem of getting *the derivative of an area* we can use this same principle.

Take Δx as the principal infinitesimal, then ΔA and Δy are of the first order, by Art. 149. ΔA differs from the rectangle $y\Delta x$

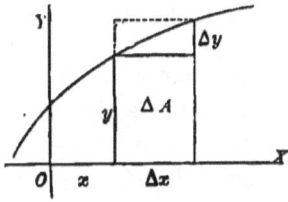

by less than the rectangle $\Delta x \Delta y$, which is of the second order, by Art. 147, Ex.; and we have

$$D_xA = \underset{\Delta x \doteq 0}{\text{limit}} \left[\frac{\Delta A}{\Delta x}\right] = \underset{\Delta x \doteq 0}{\text{limit}} \left[\frac{y\Delta x}{\Delta x}\right] = y.$$

Take the problem of *the derivative of an arc*.

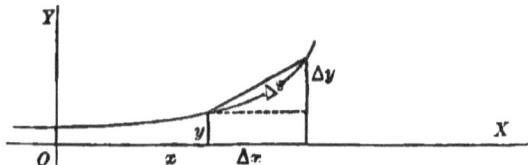

Let Δx be the principal infinitesimal; then Δs is of the first order. Δs differs from its chord $\sqrt{(\Delta x)^2 + (\Delta y)^2}$ by an infinitesimal of a higher order, by Art. 165. Hence we have

$$D_xs = \underset{\Delta x \doteq 0}{\text{limit}}\left[\frac{\Delta s}{\Delta x}\right] = \underset{\Delta x \doteq 0}{\text{limit}}\left[\frac{\sqrt{(\Delta x)^2+(\Delta y)^2}}{\Delta x}\right] = \underset{\Delta x \doteq 0}{\text{limit}}\left[\sqrt{1+\left(\frac{\Delta y}{\Delta x}\right)^2}\right],$$

$$D_xs = \sqrt{1+(D_xy)^2}.$$

178. In general,

$$D_xfx = \underset{\Delta x \doteq 0}{\text{limit}} \left[\frac{f(x+\Delta x) - fx}{\Delta x}\right];$$

therefore
$$\frac{f(x+\Delta x) - fx}{\Delta x} = D_xfx + \epsilon,$$

where ϵ is an infinitesimal, by Art. 7.

$$f(x+\Delta x) - fx = D_xfx \cdot \Delta x + \epsilon \Delta x.$$

But $f(x+\Delta x) - fx$ is the actual increment of fx, caused by the increment Δx of x. $\epsilon \Delta x$ is of as high an order as the second, if we take Δx as our principal infinitesimal; and we get the important result that $D_xfx \cdot \Delta x$ differs from the actual increment of fx by an infinitesimal of a higher order, and may consequently be used in place of Δfx in any case where we have to deal with *the limit of the ratio or of the sum of such increments*. This quantity, $D_x fx \cdot \Delta x$ is called the *differential of fx*, and is denoted by dfx, d being a symbol for the word differential.

By the definition of differential,

$$dx = D_x x \cdot \Delta x = \Delta x.$$

This definition may now be restated as follows: *The differential of the independent variable is the actual increment of that variable. The differential of a function is the derivative of the function multiplied by the differential of the independent variable;*

or formulating, $\quad dx = \Delta x,$

$$dy = D_x y \cdot dx,$$

y being a function of x.

It is to be noted that *a differential is an infinitesimal, and that it differs from an infinitesimal increment by an infinitesimal of a higher order.*

179. Since $\quad dy = D_x y \cdot dx,$

$$\frac{dy}{dx} = D_x y.$$

As, by Art. 73, $\quad D_y x = \dfrac{1}{D_x y},$

$$D_y x = \frac{dx}{dy}.$$

Consequently, if two quantities are so connected that either is a function of the other, the *derivative* of either with respect to the other is the *actual ratio of the differential of the first to the differential of the second.*

180. The differential notation has the advantage over the derivative notation, that it is apparently simpler, and that the formulas in which it is used are more symmetrical than those in which the other notation is employed; and although the differential is defined by the aid of the derivative, and the formulas

for the differentials of functions are obtained from the formulas for the derivatives of the same functions, *there is a practical advantage*, after the formulas have once been obtained, *in regarding the differential as the main thing, and looking at the derivative as the quotient of two differentials.*

181. By multiplying each of our *derivative* formulas by dx, we get the following set of formulas for the *differentials* of functions.

$$da = 0;$$

$$d(ax) = a\,dx;$$

$$d(x^n) = nx^{n-1}\,dx;$$

$$d(\log x) = \frac{dx}{x};$$

$$da^x = a^x \log a \cdot dx;$$

$$de^x = e^x\,dx;$$

$$d\sin x = \cos x \cdot dx;$$

$$d\cos x = -\sin x \cdot dx;$$

$$d\tan x = \sec^2 x \cdot dx;$$

$$d\operatorname{ctn} x = -\csc^2 x \cdot dx;$$

$$d\sec x = \sec x \tan x \cdot dx;$$

$$d\csc x = -\csc x \operatorname{ctn} x \cdot dx;$$

$$d\operatorname{vers} x = \sin x \cdot dx;$$

$$d\sin^{-1} x = \frac{dx}{\sqrt{(1-x^2)}};$$

$$d\cos^{-1} x = -\frac{dx}{\sqrt{(1-x^2)}};$$

$$d\tan^{-1} x = \frac{dx}{1+x^2};$$

$$d\cot^{-1}x = -\frac{dx}{1+x^2};$$

$$d\sec^{-1}x = \frac{dx}{x\sqrt{(x^2-1)}};$$

$$d\csc^{-1}x = -\frac{dx}{x\sqrt{(x^2-1)}};$$

$$d\operatorname{vers}^{-1}x = \frac{dx}{\sqrt{(2x-x^2)}};$$

$$d(u+v+w+\cdots) = du+dv+dw+\cdots;$$

$$d(uv) = udv + vdu;$$

$$d\frac{u}{v} = \frac{vdu-udv}{v^2};$$

$$dA = ydx;$$

$$ds = \sqrt{(dx)^2+(dy)^2}.$$

The formula $\quad D_x fy = D_y fy \cdot D_x y$

is no longer necessary, as it gives us

$$dfy = D_y fy \cdot dy = \frac{dfy}{dy}dy = dfy, \qquad \text{an identity.}$$

Examples.

Work the examples in Chap. IV. by the differential formulas just given, remembering that

$$D_x y = \frac{dy}{dx}.$$

182. The differential notation is especially convenient in dealing with problems in *integration*, and leads to an entirely new way of looking at an integral.

Let $$y = x^2,$$
and suppose that x changes from the value 1 to the value 5; to find the whole change produced in y. Let x change by successive increments, each of which may be called $\varDelta x$; then the whole change in y is the sum of the corresponding increments of y, which we will indicate by $\sum_{x=1}^{x=5} \varDelta y$. The whole change in y is the actual sum of these infinitesimal increments; it is then the limit of their sum as $\varDelta x$ is indefinitely decreased, and each $\varDelta y$ decreases correspondingly; that is, it is limit $\sum_{x=1}^{x=5} \varDelta y$. But as we are dealing with the limit of a sum of infinitesimals where the limit is, from the nature of the case, finite, each term may be replaced by any infinitesimal differing from it by an infinitesimal of a higher order (Art. 162). Each $\varDelta y$ may then be replaced by the corresponding dy, and we get as the whole change produced in the value of y,
$$\text{limit} \sum_{x=1}^{x=5} d(x^2) = \text{limit} \sum_{x=1}^{x=5} 2x\,dx.$$

As $$y = x^2,$$
this change must be
$$[x^2]_{x=5} - [x^2]_{x=1} = 25 - 1 = 24,$$
and we get the limit of the sum of a set of differentials appearing as the difference between two values of the corresponding function.

183. Suppose that in any fx we change x from x_0 to x_1 by giving to x successive increments. The whole change, $fx_1 - fx_0$, must be the sum of the partial changes produced by the increments given to x; or
$$fx_1 - fx_0 = \sum_{x=x_0}^{x=x_1} \varDelta fx.$$

If the increments given to x be indefinitely decreased in magnitude while sufficiently increased in number to still fill the gap between x_0 and x_1,

$$fx_1 - fx_0 = \lim_{\Delta x \doteq 0} \sum_{x=x_0}^{x=x_1} \Delta fx = \lim \sum_{x=x_0}^{x=x_1} dfx,$$

by Arts. 162 and 178,

$$= \lim \sum_{x=x_0}^{x=x_1} D_x fx \cdot dx.$$

Call $$D_x fx = Fx,$$

then $$fx = \int_x Fx$$

and $$\lim \sum_{x=x_0}^{x=x_1} Fx \cdot dx = [\int_x Fx]_{x=x_1} - [\int_x Fx]_{x=x_0},$$

and *the limit of the sum of a set of differentials is the difference between two values of an integral.* Such a limit is called a *definite integral*, and is indicated by $\int_{x_0}^{x_1}$, x_0 and x_1 being the values between which the sum is taken. *As a definite integral is the difference between two values of an ordinary integral, it contains no arbitrary constant.*

184. Regarding an integral as the limit of a sum gives a new meaning to some of our old formulas. Take, for example, the case of finding an area. Required the area bounded by the parabola $y^2 = 4x$, the axis of X and any ordinate y_0.

The area in question is the limit of the sum of rectangles of which $y\Delta x$ may be taken as any one, and the sum is to be taken between the values 0 and x_0 of x. We have then

$$A = \lim \sum_{x=0}^{x=x_0} y\Delta x = \lim \sum_{x=0}^{x=x_0} ydx;$$

hence $$A = \int_0^{x_0} ydx,$$

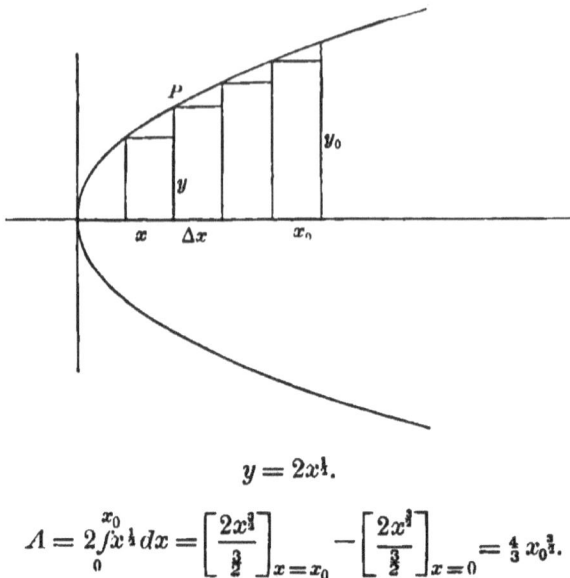

$$y = 2x^{\frac{1}{2}}.$$

$$A = 2\int_0^{x_0} x^{\frac{1}{2}} dx = \left[\frac{2x^{\frac{3}{2}}}{\frac{3}{2}}\right]_{x=x_0} - \left[\frac{2x^{\frac{3}{2}}}{\frac{3}{2}}\right]_{x=0} = \tfrac{4}{3} x_0^{\frac{3}{2}}.$$

185. We can now take up some new problems that could not be conveniently approached while the integral was treated merely as an inverse function, and we shall consider very briefly one connected with the subject of *centre of gravity*.

The centre of gravity of a body is a point so situated that the body will remain motionless in any position in which it may be placed, provided this point is supported.

Suppose a heavy plane curve, of which equal areas have equal weights, placed in a horizontal position. The tendency of any particle to produce rotation about a given axis is the weight of the particle multiplied by its distance from the axis. If the axis passes through the centre of gravity, the sum of all these tendencies must be zero, or the body would rotate.

Let us consider the centre of gravity of a segment of the parabola

$$y^2 = 2mx,$$

cut off by any double ordinate.

Suppose the parabola horizontal, and let X and Y be the coördinates of the required centre of gravity. Inscribe in the parabola small rectangles having their sides parallel to the axis of Y. The tendency of any one of these rectangles, as AB, to produce rotation about the ordinate through the centre of gravity, is its

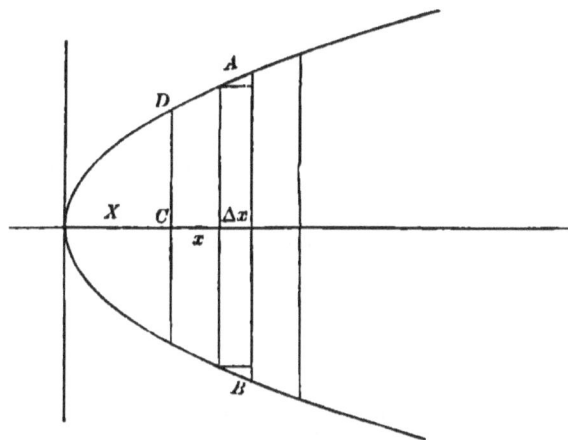

weight, which may be represented by its area, $2y\,\varDelta x$, multiplied by its distance from the ordinate in question. If the rectangle were so narrow that we could regard its weight as concentrated along its nearest side, this distance would be $(x-X)$; and if we decrease $\varDelta x$ indefinitely, the required distance will approach this as its limit.

The tendency of this rectangle to produce rotation is then, roughly, $2y(x-X)\varDelta x$; and the smaller the value of $\varDelta x$, the nearer this comes to being an *exact* expression. The tendency of all the rectangles is $\sum_{x=0}^{x=x_1} 2y(x-X)\varDelta x$. The smaller the rectangles, the nearer their sum comes to the whole area of the curve, and we shall have as the tendency of the whole curve to rotate about CD $\lim \sum_{x=0}^{x=x_1} 2y(x-X)\varDelta x$ or $\int_0^{x_1} 2y(x-X)dx$; but as CD passes through the centre of gravity, this must equal zero.

$$\int_{t}^{x_1} 2y(x-X)dx = 0.$$

$$y = \sqrt{(2mx)};$$

hence
$$2\int_0^{x_1} \sqrt{(2mx)}\,(x-X)\,dx = 0,$$

$$\int_0^{x_1} x^{\tfrac{3}{2}}\,dx = X\int_0^{x_1} x^{\tfrac{1}{2}}\,dx,$$

$$\tfrac{2}{5} x_1^{\tfrac{5}{2}} = \tfrac{2}{3} X x_1^{\tfrac{3}{2}},$$

$$X = \tfrac{3}{5} x_1.$$

By similar reasoning, we find, as the tendency to rotate about a line through the centre of gravity and parallel to the axis of X, $\int_{-y_1}^{y_1}(x_1-x)(y-Y)\,dy$. This must equal zero.

$$\int_{-y_1}^{y_1}\left(x_1 - \frac{y^2}{2m}\right)(y-Y)\,dy = 0;$$

$$\int_{-y_1}^{y_1}\left(x_1 y - x_1 Y - \frac{y^3}{2m} + \frac{y^2 Y}{2m}\right)dy = 0;$$

$$\left[x_1\frac{y^2}{2} - x_1 Y y - \frac{y^4}{8m} + \frac{y^3}{6m} Y\right]_{y=-y_1}^{y=y_1} = 0;$$

$$\frac{x_1 y_1^2}{2} - x_1 Y y_1 - \frac{y_1^4}{8m} + \frac{y_1^3 Y}{6m} - \frac{x_1 y_1^2}{2} - x_1 Y y_1 + \frac{y_1^4}{8m} + \frac{y_1^3}{6m} Y = 0;$$

$$\left(\frac{y_1^3}{3m} - 2 x_1 y_1\right) Y = 0;$$

$$Y = 0;$$

and $(\tfrac{3}{5} x_1, 0)$ is the required centre of gravity.

Differentials of Different Orders.

186. As the differential of a function is by definition a new function of the independent variable, we may deal with its differential.

$d(dy)$ is called the *second differential* of y, and is denoted by $d^2 y$; $d(d^2 y)$ is called the *third differential* of y, and is denoted by $d^3 y$; and so on. $\quad d(d^{n-1} y) = d^n y.$

In dealing with differentials of a higher order than the first, it is customary to make the assumption that the *differential*, that is, the increment (Art. 178), *of the independent variable is constant*, since this assumption greatly simplifies the results, and is always allowable when the variable in question is really independent, as we can then suppose it to change by equal increments.

187. Making the assumption that the *differential of the independent variable is constant*, we have very simple relations between differentials and derivatives of different orders.

By Art. 178, $\quad dy = D_x y \cdot dx$,

then $\quad d^2 y = d(dy) = D_x dy \cdot dx = D_x(D_x y \cdot dx) dx = D_x^2 y \cdot dx^2$,

as dx is a constant. It can be shown in the same way that

$$d^3 y = D_x^3 y \cdot dx^3,$$

and that $\quad d^n y = D_x^n y \cdot dx^n$.

It will be noticed that when dx is the principal infinitesimal, $d^n y$ is an *infinitesimal of the nth order*.

From the results just obtained, we get,

$$D_x^2 y = \frac{d^2 y}{dx^2},$$

$$D_x^3 y = \frac{d^3 y}{dx^3},$$

$$D_x^n y = \frac{d^n y}{dx^n},$$

and the differential notation is generally used in place of the derivative, even in the case of derivatives of higher order than the first; but in using $\frac{d^n y}{dx^n}$ for $D_x^n y$, it must be kept in mind that *the two expressions are equivalent only when* x *is the inde-*

pendent variable. If, for example, x were a function of a third variable, and were compelled to change in some particular way, we could no longer assume that dx was constant, and the differential expressions for the derivatives would be much more complicated.

188. Let us work out the second derivative of y *without any assumption* as to the value of dx.

$$D_x^2 y = \frac{dD_x y}{dx} = \frac{d\frac{dy}{dx}}{dx} = \frac{dx\, d^2y - dy\, d^2x}{dx^3},$$

since $\dfrac{dy}{dx}$ is an ordinary fraction, and its differential can be found by the formula

$$d\frac{u}{v} = \frac{v\, du - u\, dv}{v^2}.$$

Examples.

(1) Show that

$$D_x^3 y = \frac{d^3y\, dx^2 - dx\, dy\, d^3x - 3\, dx\, d^2y\, d^2x + 3\, dy\, d^2x^2}{dx^5}.$$

(2) If $\qquad y = \log z,$

find d^2y, d^3y, and d^4y, assuming that z is the independent variable, and again making no assumption concerning z. Compare your last results with those obtained by letting $z = \sin x$, and taking x as the independent variable.

189. In using differentials of higher order than the first, if the assumption is made that the differential of the independent variable is constant, it is better to indicate this by preserving the derivative form, even when using the differential notation. Take, for example, the formula for the radius of curvature of a plane curve,

$$\rho = -\frac{[1 + (D_x y)^2]^{\frac{3}{2}}}{D_x^2 y}.$$

We should write it
$$\rho = -\frac{\left[1+\left(\frac{dy}{dx}\right)^2\right]^{\frac{3}{2}}}{\frac{d^2y}{dx^2}},$$

and not
$$\rho = -\frac{(dx^2+dy^2)^{\frac{3}{2}}}{dx \cdot d^2y},$$

if we wished to indicate that x was the independent variable. If we make no such assumption, we must substitute for $D_x^2 y$ the value given in Art. 188, and we can then reduce the formula to

$$\rho = -\frac{[dx^2+dy^2]^{\frac{3}{2}}}{dxd^2y - dyd^2x}.$$

190. The subject of *differentials of different orders* is closely connected with that of *finite differences* or *increments of different orders*.

If y is a function of x, and any fixed increment Δx is given to x, there will be produced a corresponding increment Δy in the value of y; Δy, however, is not a fixed value, but varies with the value of x considered. For example, if

$$y = x^3,$$
$$\Delta y = 3x^2 \Delta x + 3x(\Delta x)^2 + (\Delta x)^3,$$

and is obviously a function of x, and therefore will be changed by changing x. The change produced in Δy by giving x another increment, Δx, is called the second increment of y, and is indicated by $\Delta^2 y$, and is a new function of x. The increment of the second increment is the third increment $\Delta^3 y$, and so on; and in general
$$\Delta(\Delta^{n-1}y) = \Delta^n y.$$

If
$$y = x^3,$$
$$\Delta^2 y = 6x(\Delta x)^2 + 6(\Delta x)^3.$$

The whole change produced in a function by giving several equal increments to the variable can be neatly expressed in terms of successive increments.

$$y = fx,$$

$$f(x + \Delta x) = y + \Delta y.$$

Add Δx again, y becomes $y + \Delta y$, Δy becomes $\Delta y + \Delta^2 y$, and we have

$$f(x + 2\Delta x) = y + 2\Delta y + \Delta^2 y.$$

Repeat the operation, y becomes $y + \Delta y$, $2\Delta y$ becomes $2(\Delta y + \Delta^2 y)$, $\Delta^2 y$ becomes $\Delta^2 y + \Delta^3 y$, and we have

$$f(x + 3\Delta x) = y + 3\Delta y + 3\Delta^2 y + \Delta^3 y.$$

In like manner,

$$f(x + 4\Delta x) = y + 4\Delta y + 6\Delta^2 y + 4\Delta^3 y + \Delta^4 y.$$

Example.

Show that, if

$$f[x + (n-1)\Delta x] = y + (n-1)\Delta y + \frac{(n-1)(n-2)}{2!}\Delta^2 y$$

$$+ \frac{(n-1)(n-2)(n-3)}{3!}\Delta^3 y + \cdots,$$

$$f(x + n\Delta x) = y + n\Delta y + \frac{n(n-1)}{2!}\Delta^2 y + \frac{n(n-1)(n-2)}{3!}\Delta^3 y + \cdots;$$

and that, consequently, the second formula always holds.

191. If Δx is infinitesimal, we have seen that dy differs from Δy by an infinitesimal of a higher order, and therefore may be used instead of Δy in all cases where we are dealing with the limit of a ratio or of a sum of such increments. The same relation holds between $d^2 y$ and $\Delta^2 y$, and in general between $d^n y$ and $\Delta^n y$, as we can prove by the aid of the following lemma.

Lemma.

192. *If a function of* x *contains besides* x *a letter* a, *which is independent of* x, *and becomes zero when* a *is zero, no matter what the value of* x, *its derivative with respect to* x *will be zero when* a *is zero.*

For, since a, being independent of x, is treated as a constant during the operation of differentiation, it can make no difference in the result whether we give it any particular value before or after that operation. But if we give a the value zero before we differentiate, our function by hypothesis is equal to zero, and is therefore constant, and its derivative is zero. Hence the lemma.

It follows that, if the function is infinitesimal when a *is infinitesimal, whatever the value of* x, *its derivative with respect to* x *will also be infinitesimal when* a *is infinitesimal.*

As an example, consider the function $\log(1+ax)$, which equals zero when $a = 0$.

$$\frac{d\log(1+ax)}{dx} = \frac{a}{1+ax} = 0 \text{ when } a = 0.$$

193. Let Δx be infinitesimal. Then, by Art. 178,

$$\frac{\Delta y}{\Delta x} = D_x y + \epsilon,$$

where ϵ approaches zero as $\Delta x \doteq 0$. Increase x by Δx, and the increments of the two members of the equation will be equal.

$$\frac{\Delta^2 y}{\Delta x} = \Delta(D_x y) + \Delta \epsilon.$$

Divide by Δx:
$$\frac{\Delta^2 y}{(\Delta x)^2} = \frac{\Delta(D_x y)}{\Delta x} + \frac{\Delta \epsilon}{\Delta x},$$

$$\lim_{\Delta x \doteq 0}\left[\frac{\Delta^2 y}{(\Delta x)^2}\right] = D_x^2 y + \lim_{\Delta x \doteq 0}\left[\frac{\Delta \epsilon}{\Delta x}\right];$$

but, by Art. 192, $\quad\underset{\Delta x \doteq 0}{\text{limit}} \left[\dfrac{\Delta \epsilon}{\Delta x}\right] = 0.$

Hence, $\quad\underset{\Delta x \doteq 0}{\text{limit}} \left[\dfrac{\Delta^2 y}{(\Delta x)^2}\right] = D_x^2 y = \dfrac{d^2 y}{dx^2},\quad$ by Art. 187.

$$\dfrac{\Delta^2 y}{(\Delta x)^2} = \dfrac{d^2 y}{dx^2} + a,$$

where a is infinitesimal, by Art. 7.

But $\quad\quad\quad\quad\quad\quad\quad \Delta x = dx \quad\quad\quad\quad\quad$ (Art. 178);

hence $\quad\quad\quad\quad\quad\quad \dfrac{\Delta^2 y}{dx^2} = \dfrac{d^2 y}{dx^2} + a,$

$$\Delta^2 y = d^2 y + a\,dx^2.$$

$d^2 y$ is an infinitesimal of the second order, by Art. 187. $a\,dx^2$ is of the third order; consequently, $d^2 y$ may be used in place of $\Delta^2 y$ in problems concerning the limit of a ratio or of a sum.

By similar reasoning, it can be shown that

$$\Delta^3 y = d^3 y + a\,dx^3;$$

and, in general, that $\quad \Delta^n y = d^n y + a\,dx^n,$

when Δx is infinitesimal.

CHAPTER XII.

FUNCTIONS OF MORE THAN ONE VARIABLE.

Partial Derivatives.

194. Up to this time we have considered only functions of a single variable, but a complete treatment of our subject requires us to study functions of two or more *independent* variables.

Plane Analytic Geometry has furnished us with numerous examples of functions of the former kind; *Analytic Geometry of Three Dimensions* introduces us to functions of the latter sort.

The equation of a surface contains three variables, x, y, and z, and any one may be expressed as a function of the other two; and when this is done, the one so expressed may be changed by changing either of the others, or by changing them both, as they are entirely independent.

195. The derivative of a function of several variables obtained on the hypothesis that only one of them changes, is called a *partial derivative;* and, as all the variables except one are, for the time being, treated as constants, a *partial derivative* can be obtained by the rules for differentiating a function of one variable.

For example: $\quad D_x x^2 y = 2xy, \quad$ if x alone changes;

$$D_y x^2 y = x^2, \quad \text{if } y \text{ alone changes.}$$

$2xy$ is the partial derivative of $x^2 y$ with respect to x, and x^2 is the partial derivative of $x^2 y$ with respect to y.

We shall represent partial derivatives by our old derivative notation, indicating ordinary or complete derivatives, when it is necessary to make any distinction between the two, by the ratio of two differentials.

196. If a function contains two variables, its partial derivative with respect to either will generally contain both variables, and may be differentiated again with respect to either of them.

Take $x^2 y^2$.

$$D_x x^2 y^2 = 2xy^2;$$

$$D_x^2 x^2 y^2 = 2y^2;$$

$$D_y D_x x^2 y^2 = 4xy;$$

$$D_y x^2 y^2 = 2x^2 y;$$

$$D_x D_y x^2 y^2 = 4xy;$$

$$D_y^2 x^2 y^2 = 2x^2.$$

Take $u = x \log y$.

$$D_x u = \log y;$$

$$D_x^2 u = 0;$$

$$D_y D_x u = \frac{1}{y};$$

$$D_y^2 D_x u = -\frac{1}{y^2};$$

$$D_y u = \frac{x}{y};$$

$$D_x D_y u = \frac{1}{y};$$

$$D_y^2 u = -\frac{x}{y^2};$$

$$D_x D_y^2 u = -\frac{1}{y^2}.$$

197. In both these examples we see that $D_x D_y u$ is the same as $D_y D_x u$, and in the second

$$D_y^2 D_x u = D_x D_y^2 u.$$

Let us see whether it is true in general that the order in which the differentiations are performed is immaterial.

Let $u = f(x,y)$.

To see if $D_y D_x u = D_x D_y u$.

$$D_y u = \lim_{\Delta y \doteq 0} \left[\frac{f(x, y + \Delta y) - f(x, y)}{\Delta y} \right] = \frac{f(x, y + \Delta y) - f(x, y)}{\Delta y} + \varepsilon,$$

by Art. 7, where ε is an infinitesimal and a function of x, y, and Δy. Similarly, $D_x u = \dfrac{f(x + \Delta x, y) - f(x, y)}{\Delta x} + \varepsilon'$,

where ε' is an infinitesimal and a function of x, y, and Δx.

$D_x D_y u$ is equal to

$$\lim_{\Delta x \doteq 0} \left[\frac{f(x + \Delta x, y + \Delta y) - f(x + \Delta x, y) - f(x, y + \Delta y) + f(x, y)}{\Delta x \, \Delta y} \right]$$
$$+ D_x \varepsilon; \quad [1]$$

$D_y D_x u$ is equal to

$$\lim_{\Delta y \doteq 0} \left[\frac{f(x + \Delta x, y + \Delta y) - f(x, y + \Delta y) - f(x + \Delta x, y) + f(x, y)}{\Delta x \, \Delta y} \right]$$
$$+ D_y \varepsilon'. \quad [2]$$

The second expression for $D_y u$ is absolutely true, whatever the value of Δy, and so is the expression for $D_x D_y u$. We may then suppose Δy to approach indefinitely near zero, and $D_x D_y u$ will be equal to the limiting value approached by the second member of [1]. The limit of ε as Δy approaches 0 is 0; therefore, by Art. 192,
$$\lim_{\Delta y \doteq 0} [D_x \varepsilon] = 0,$$

and $D_x D_y u$ is equal to

$$\lim \left[\frac{f(x + \Delta x, y + \Delta y) - f(x + \Delta x, y) - f(x, y + \Delta y) + f(x, y)}{\Delta x \, \Delta y} \right],$$

as both Δy and Δx approach 0.

By similar reasoning, it may be shown that $D_y D_x u$ is this same limit, and hence that $D_x D_y u = D_y D_x u$.

By applying this theorem at each step, we may prove that, *in obtaining any successive partial derivatives, the order in which the differentiations occur is of no consequence.*

For example, let us show that

$$D_x^2 D_y u = D_y D_x^2 u\,;$$

$$D_x^2 D_y u = D_x(D_x D_y u) = D_x(D_y D_x u) = D_x D_y D_x u$$

$$= D_y D_x D_x u = D_y D_x^2 u.$$

198. In a previous chapter, we saw that, while the *increment* of a function due to any increment of the variable is generally a very complex expression, the *differential* of the function, which differs from the true increment only by an infinitesimal of a higher order than the increment of function or variable when the latter is infinitesimal, is usually very much simpler, and yet can be used instead of the true increment in many important problems.

It is worth while to see if we cannot get a simple expression capable of replacing the infinitesimal increment of a function of two or more variables in similar problems.

A function of two independent variables may be changed by changing either of the variables alone, or by changing both.

Suppose we give to each variable an infinitesimal increment of the same order. Let $u = f(x,y)$.

Increase x by Δx and y by Δy,

$$\Delta u = f(x + \Delta x, y + \Delta y) - f(x,y).$$

Add and subtract $f(x, y + \Delta y)$, and we get

$$\Delta u = f(x + \Delta x, y + \Delta y) - f(x, y + \Delta y) + f(x, y + \Delta y) - f(x,y).$$

$f(x, y + \Delta y) - f(x,y)$ is the increment of $f(x,y)$ produced by changing y alone, and differs from $D_y f(x,y) \Delta y$ by an infinitesimal of a higher order than Δy, by Art. 178. In like manner, we see that $f(x + \Delta x, y + \Delta y) - f(x, y + \Delta y)$ differs from $D_x f(x, y + \Delta y) \Delta x$ by an infinitesimal of a higher order than Δx.

$D_x f(x, y + \Delta y)$ is a new function of x and y, and any infinitesimal change in y will produce in it a change of the same order, by Art. 149. $D_x f(x, y + \Delta y)$, then, differs from $D_x f(x,y)$ by an infinitesimal of the same order as Δy, and $D_x f(x, y + \Delta y) \Delta x$ differs from $D_x f(x,y) \Delta x$ by an infinitesimal of the second order.

$D_x f(x,y) \Delta x + D_y f(x,y) \Delta y$, or, using the differential notation and remembering that x and y are both independent, $D_x f(x,y) dx + D_y f(x,y) dy$ differs from the true increment of u by an infinitesimal of a higher order than dx and dy, and therefore may be used in place of Δu whenever the limit of a ratio or the limit of a sum is sought. This is called the complete differential of u, and is indicated by du; hence, when

$$u = f(x,y),$$

$$du = D_x u\, dx + D_y u\, dy.$$

Example.

Prove that, if $\quad u = f(x,y,z),$

$$du = D_x u\, dx + D_y u\, dy + D_z u\, dz.$$

199. Partial derivatives may very often be used with profit in obtaining ordinary or complete derivatives. Suppose that

$$y = Fx \text{ and } z = F_1 x \text{ and } u = f(y,z);$$

u is indirectly a function of x, and we can therefore speak of the complete derivative of u with respect to x, which we shall indicate by $\dfrac{du}{dx}$.

We wish to find the limit of the ratio $\frac{\Delta u}{\Delta x}$. In so doing, we can replace Δu by du, which equals $D_y u \Delta y + D_z u \Delta z$, since, as y and z are not independent variables, Δy and Δz differ from dy and dz;

hence
$$\frac{du}{dx} = \underset{\Delta x \doteq 0}{\text{limit}} \left[D_y u \frac{\Delta y}{\Delta x} + D_z u \frac{\Delta z}{\Delta x} \right],$$

or
$$\frac{du}{dx} = D_y u \frac{dy}{dx} + D_z u \frac{dz}{dx}.$$

EXAMPLE.

To find $\dfrac{d \sin(y^2 - z)}{dx}$, knowing that $\begin{cases} y = \log x, \\ z = x^2, \end{cases}$

Solution: $\quad D_y \sin(y^2 - z) = 2y \cos(y^2 - z),$

$$D_z \sin(y^2 - z) = -\cos(y^2 - z).$$

$$\frac{dy}{dx} = \frac{1}{x},$$

$$\frac{dz}{dx} = 2x,$$

$$\frac{d \sin(y^2 - z)}{dx} = \frac{2y \cos(y^2 - z)}{x} - 2x \cos(y^2 - z)$$

$$= \frac{2(y - x^2) \cos(y^2 - z)}{x}.$$

Confirm this result by expressing y and z in terms of x before differentiating.

200. If $\quad u = f(x, y)$ and $y = Fx$,

the formula of the last article becomes

$$\frac{du}{dx} = D_x u + D_y u \frac{dy}{dx}.$$

EXAMPLES.

(1) $\left.\begin{array}{l} u = z^2 + y^3 + zy \\ z = \sin x \\ y = e^x \end{array}\right\}$ Find $\dfrac{du}{dx}$.

Ans. $\dfrac{du}{dx} = (3y^2 + z)e^x + (2z + y)\cos x$.

(2) $\left.\begin{array}{l} u = \log\dfrac{x}{y} \\ y = \sin x \end{array}\right\}$ Find $\dfrac{du}{dx}$.
Ans. $\dfrac{du}{dx} = \dfrac{1}{x} - \operatorname{ctn} x$.

(3) $\left.\begin{array}{l} u = \tan^{-1}(xy) \\ y = e^x \end{array}\right\}$ Find $\dfrac{du}{dx}$.
Ans. $\dfrac{du}{dx} = \dfrac{e^x x + y}{1 + x^2 y^2}$.

(4) $u = \sin^{-1}\left(\dfrac{z}{y}\right)$ when z and y are functions of x. Find $\dfrac{du}{dx}$.

(5) $u = \sqrt{\left(\dfrac{z^2 - y^2}{z^2 + y^2}\right)}$ when z and y are functions of x. Find $\dfrac{du}{dx}$.

201. Higher derivatives of a function of functions of x can be obtained by an easy application of the method suggested by the formulas above.

For example:
$$u = f(y, z),$$
$$y = Fx,$$
$$z = F_1 x,$$

required $\dfrac{d^2 u}{dx^2}$.

$$\dfrac{d^2 u}{dx^2} = \dfrac{d\left(\dfrac{du}{dx}\right)}{dx} = \left[D_y^2 u \dfrac{dy}{dx} + D_y D_z u \dfrac{dz}{dx}\right]\dfrac{dy}{dx}$$

$$+ \left[D_y D_z u \dfrac{dy}{dx} + D_z^2 u \dfrac{dz}{dx}\right]\dfrac{dz}{dx} + \left[D_y u \dfrac{d^2 y}{dx^2} + D_z u \dfrac{d^2 z}{dx^2}\right]$$

$$= D_y^2 u \left(\dfrac{dy}{dx}\right)^2 + 2 D_y D_z u \dfrac{dy}{dx} \cdot \dfrac{dz}{dx} + D_z^2 u \left(\dfrac{dz}{dx}\right)^2 + D_y u \dfrac{d^2 y}{dx^2} + D_z u \dfrac{d^2 z}{dx^2}.$$

In obtaining this formula, since y and z are given functions of x, $\dfrac{dy}{dx}$ and $\dfrac{dz}{dx}$ are also explicit functions of x, and are therefore treated as constants in obtaining the partial derivatives with respect to y and z; but now $\dfrac{du}{dx}$ is a function of $(x, y$ and $z)$, hence we must take also its partial derivative with respect to x.

EXAMPLE.

Given
$$u = f(x,y),$$
$$y = Fx,$$

obtain $\dfrac{d^2 u}{dx^2}$ and $\dfrac{d^3 u}{dx^3}$.

Implicit Functions.

202. If, instead of having y given in terms of x, we have an equation connecting x and y, y is called an implicit function of x, and $\dfrac{dy}{dx}$ can be readily found by the aid of Partial Derivatives.

Suppose
$$f(x,y) = 0,$$
to find $\dfrac{dy}{dx}$. Call $f(x,y)$ u.

Then
$$u = 0;$$

hence $\dfrac{du}{dx}$ must also equal zero,

$$\frac{du}{dx} = D_x u + D_y u \frac{dy}{dx} = 0,$$

$$\frac{dy}{dx} = -\frac{D_x u}{D_y u}.$$

EXAMPLES.

(1) $ax^m - y e^{ny} = 0$. Find $\dfrac{dy}{dx}$.

Solution:
$$D_x u = m a x^{m-1},$$
$$D_y u = -e^{ny} - nye^{ny},$$
$$\frac{dy}{dx} = \frac{-D_x u}{D_y u} = \frac{max^{m-1}}{(1+ny)e^{ny}};$$

or, as
$$ax^m = ye^{ny},$$
$$max^{m-1} = \frac{mye^{ny}}{x},$$

and
$$\frac{dy}{dx} = \frac{my}{(1+ny)x}.$$

(2) $\frac{x^2}{a^2} + \frac{y^2}{b^2} - 1 = 0$. Find $\frac{dy}{dx}$. Ans. $\frac{dy}{dx} = -\frac{b^2 x}{a^2 y}$.

(3) $x^y - y^x = 0$. Find $\frac{dy}{dx}$. Ans. $\frac{dy}{dx} = \frac{y^2 - xy \log y}{x^2 - xy \log x}$.

(4) $\sin(xy) - mx = 0$. Find $\frac{dy}{dx}$.

(5) $\left.\begin{array}{l} u^2 + x^2 + y^2 + z^2 = c^2 \\ \log(xy) + \dfrac{y}{x} = a^2 \\ \log\left(\dfrac{z}{x}\right) + zx = b^2 \end{array}\right\}$ Find $\dfrac{du}{dx}$.

Ans. $\dfrac{du}{dx} = \dfrac{1}{u}\left(\dfrac{y^2(x-y)}{x(x+y)} + \dfrac{z^2(xz-1)}{x(xz+1)} - x\right).$

203. We can get $\dfrac{d^2 y}{dx^2}$ by the aid of the formula of Art. 201, remembering that
$$\frac{d^2 u}{dx^2} = 0.$$
$$\frac{d^2 u}{dx^2} = D_x^2 u + 2 D_x D_y u \frac{dy}{dx} + D_y^2 u \left(\frac{dy}{dx}\right)^2 + D_y u \frac{d^2 y}{dx^2} = 0,$$

$$\frac{dy}{dx} = -\frac{D_x u}{D_y u},$$

$$D_x^2 u - 2\left(D_x D_y u \frac{D_x u}{D_y u}\right) + D_y^2 u \left(\frac{D_x u}{D_y u}\right)^2 + D_y u \frac{d^2 y}{dx^2} = 0,$$

$$\frac{d^2 y}{dx^2} = -\frac{D_x^2 u (D_y u)^2 - 2 D_x D_y u \, D_x u \, D_y u + D_y^2 u (D_x u)^2}{(D_y u)^3}.$$

EXAMPLES.

(1) $y^3 + x^3 - 3axy = 0$. Find $\dfrac{d^2 y}{dx^2}$.   Ans. $\dfrac{d^2 y}{dx^2} = -\dfrac{2a^3 xy}{(y^2 - ax)^3}$.

(2) $x^4 + 2ax^2 y - ay^3 = 0$. Find $\dfrac{dy}{dx}$ and $\dfrac{d^2 y}{dx^2}$.

CHAPTER XIII.

CHANGE OF VARIABLE.

204. If we use the differential notation, we have seen that there is no need of distinguishing carefully between function and independent variable, a single formula always giving a relation between the two differentials by which either can be expressed in terms of the other. This, however, is the case only when we are dealing with differentials of the first order. A differential of the second order or of a higher order has been defined by the aid of a derivative, which always implies the distinction between function and variable, and on the hypothesis of an important difference in the natures of the increments of function and variable; namely, that the increment of the independent variable is a constant magnitude, and that, consequently, its derivative and differential are zero.

If, in any function involving differentials of a higher order than the first, we have occasion to change the independent variable, we can no longer assume that the differential of the old independent variable is constant, but must go back and replace all the differentials of higher order than the first by values obtained on the supposition that all the differentials are variable, before we attempt the introduction of the new variable, *vide* Arts. 187 and 188.

205. In any particular example in which it is necessary to change the variable, the method just described can be easily applied.

Take the differential equation,

$$x^2 \frac{d^2u}{dx^2} + x \frac{du}{dx} + u = 0,$$

where x is the independent variable, and introduce y in place of x. Given
$$y = \log x.$$

Our d^2u here is the second differential of u taken on the assumption that x is the independent variable, and this can be indicated by writing it $d_x^2 u$, and we have

$$d_x^2 u = D_x^2 u\, dx^2 = \frac{dx\, d^2 u - du\, d^2 x}{dx}, \quad \text{by Art. 188.}$$

$$dy = \frac{dx}{x},$$

$$dx = x\, dy,$$

$$d^2 x = d(x\, dy) = x\, d^2 y + dx\, dy;$$

but
$$d^2 y = 0,$$

as y is to be the independent variable,

hence
$$d^2 x = dx\, dy,$$

and
$$d_x^2 u = \frac{x\, dy\, d^2 u - x\, du\, dy^2}{x\, dy} = d^2 u - du\, dy;$$

$$x^2 \frac{d^2 u}{dx^2} = \frac{d^2 u}{dy^2} - \frac{du}{dy},$$

$$x \frac{du}{dx} = \frac{du}{dy};$$

hence we have
$$\frac{d^2 u}{dy^2} + u = 0.$$

Examples.

(1) Change the variable from x to t in

$$\frac{d^2y}{dx^2} - \frac{x}{1-x^2}\frac{dy}{dx} + \frac{y}{1-x^2} = 0.$$

Given $\qquad x = \cos t.$

$\qquad\qquad\qquad\qquad$ Ans. $\dfrac{d^2y}{dt^2} + y = 0.$

(2) Change the variable from x to θ in the equation,

$$\frac{d^2y}{dx^2} + \frac{2x}{1+x^2}\frac{dy}{dx} + \frac{y}{(1+x^2)^2} = 0.$$

Given $\qquad \theta = \tan^{-1} x.$

$\qquad\qquad\qquad\qquad$ Ans. $\dfrac{d^2y}{d\theta^2} + y = 0.$

(3) Change from x to t in

$$(1-x^2)\frac{d^2y}{dx^2} - x\frac{dy}{dx} = 0.$$

Given $\qquad x = \cos t.$

$\qquad\qquad\qquad\qquad$ Ans. $\dfrac{d^2y}{dt^2} = 0.$

206. It is often desirable to change both variables simultaneously, and the principles already explained and illustrated apply perfectly to this case. As an example, let us see what our old expression for the radius of curvature of a plane curve becomes when we change from rectangular to polar coördinates.

Here we have
$$\left. \begin{array}{l} x = r\cos\varphi \\ y = r\sin\varphi \end{array} \right\}$$

and we shall regard φ as the new independent variable. We know that, if ρ is the radius of curvature,

$$\rho = -\frac{[1+(D_x y)^2]^{\frac{3}{2}}}{D_x^2 y}.$$

We have seen, in Art. 189, that this may be written

$$\rho = \frac{-(dx^2 + dy^2)^{\frac{3}{2}}}{dx d_x^2 y};$$

or, better still,
$$\rho = -\frac{(dx^2 + dy^2)^{\frac{3}{2}}}{dx d^2 y - dy d^2 x}.$$

$$dx = -r\sin\varphi\, d\varphi + \cos\varphi\, dr,$$

$$dy = r\cos\varphi\, d\varphi + \sin\varphi\, dr.$$

Since $d\varphi$ is constant,

$$d^2 x = -r\cos\varphi\, d\varphi^2 - 2\sin\varphi\, dr d\varphi + \cos\varphi\, d^2 r,$$

$$d^2 y = -r\sin\varphi\, d\varphi^2 + 2\cos\varphi\, dr d\varphi + \sin\varphi\, d^2 r,$$

$$(dx^2 + dy^2)^{\frac{3}{2}} = (r^2 d\varphi^2 + dr^2)^{\frac{3}{2}},$$

$$dx d^2 y - dy d^2 x = r^2 d\varphi^3 - r d\varphi d^2 r + 2 dr^2 d\varphi.$$

$$\rho = -\frac{(r^2 d\varphi^2 + dr^2)^{\frac{3}{2}}}{r^2 d\varphi^3 - r d\varphi d^2 r + 2 dr^2 d\varphi};$$

divide numerator and denominator by $d\varphi^3$,

$$\rho = \frac{-\left[r^2 + \left(\dfrac{dr}{d\varphi}\right)^2\right]^{\frac{3}{2}}}{r^2 - r\dfrac{d^2 r}{d\varphi^2} + 2\left(\dfrac{dr}{d\varphi}\right)^2}.$$

Example.

Find the radius of curvature of the circle $r = \cos\varphi$.

Ans. $\rho = \tfrac{1}{2}$.

207. A very simple example of change of variable is the following. Obtain the value of $\tan \tau$ when polar coördinates are used.

$$\tan \tau = \frac{dy}{dx}.$$

$$x = r \cos \varphi,$$

$$y = r \sin \varphi,$$

$$\frac{dy}{dx} = \frac{r \cos \varphi \, d\varphi + \sin \varphi \, dr}{-r \sin \varphi \, d\varphi + \cos \varphi \, dr}.$$

A much simpler expression can be obtained for the angle made by the tangent with the radius vector, which we shall call ϵ.

$$\epsilon = \tau - \varphi.$$

$$\tan \epsilon = \tan(\tau - \varphi) = \frac{\tan \tau - \tan \varphi}{1 + \tan \tau \tan \varphi},$$

$$\tan \tau - \tan \varphi = \frac{r \sec \varphi \, d\varphi}{-r \sin \varphi \, d\varphi + \cos \varphi \, dr},$$

$$1 + \tan \tau \tan \varphi = \frac{\sec \varphi \, dr}{-r \sin \varphi \, d\varphi + \cos \varphi \, dr}.$$

$$\tan \epsilon = \frac{r \, d\varphi}{dr}.$$

EXAMPLES.

(1) Obtain this value for $\tan \epsilon$ from a figure by the aid of infinitesimals.

(2) If
$$\left. \begin{array}{l} x = r\cos\varphi \\ y = r\sin\varphi \end{array} \right\}$$

show that
$$\frac{x + y\dfrac{dy}{dx}}{x\dfrac{dy}{dx} - y} = \frac{dr}{r\,d\varphi},$$

and that $ds^2 = dx^2 + dy^2$

becomes $ds^2 = dr^2 + r^2 d\varphi^2.$

Prove this last result from a figure.

(3) If
$$\left. \begin{array}{l} x = a(1 - \cos t) \\ y = a(nt + \sin t) \end{array} \right\}$$

express $\dfrac{d^2y}{dx^2}$ in terms of t.   Ans. $\dfrac{d^2y}{dx^2} = -\dfrac{n\cos t + 1}{a\sin^3 t}.$

(4) Given
$$\left. \begin{array}{l} x = a\cos\varphi \\ y = b\sin\varphi \end{array} \right\}$$

express $-\dfrac{\left[1 + \left(\dfrac{dy}{dx}\right)^2\right]^{\frac{3}{2}}}{\dfrac{d^2y}{dx^2}}$ in terms of φ.

Ans. $\dfrac{(a^2\sin^2\varphi + b^2\cos^2\varphi)^{\frac{3}{2}}}{ab}.$

208. The subject of change of variable can be easily treated, by the aid of the principles established in Art. 88, without introducin the idea of differentials.

$$D_z y = \frac{D_x y}{D_x x},$$

$$D_z^2 y = D_z D_z y = \frac{D_z D_z y}{D_z x} = \frac{D_z\left(\dfrac{D_z y}{D_z x}\right)}{D_z x}.$$

$$D_z\left(\frac{D_z y}{D_z x}\right) = \frac{D_z x\, D_z^2 y - D_z y\, D_z^2 x}{(D_z x)^2},$$

$$D_x^2 y = \frac{D_z x\, D_z^2 y - D_z y\, D_z^2 x}{(D_z x)^3}.$$

If x and y are given in terms of z, we can calculate the values of $D_z x$, $D_z y$, $D_z^2 x$, and $D_z^2 y$, and substitute them in these formulas. Take Example (3), Art. 205.

$$x = \cos t,$$

$$(1-x^2)D_x^2 y - x\, D_x y = 0,$$

$$D_x y = \frac{D_t y}{D_t x},$$

$$D_x^2 y = \frac{D_t x\, D_t^2 y - D_t y\, D_t^2 x}{(D_t x)^3},$$

$$D_t x = -\sin t,$$

$$D_t^2 x = -\cos t,$$

$$D_x y = -\frac{D_t y}{\sin t},$$

$$D_x^2 y = \frac{-\sin t\, D_t^2 y + \cos t\, D_t y}{-\sin^3 t},$$

$$1 - x^2 = \sin^2 t,$$

$$\sin^2 t\, \frac{-\sin t\, D_t^2 y + \cos t\, D_t y}{-\sin^3 t} + \frac{\cos t\, D_t y}{\sin t} = 0,$$

$$D_t^2 y = 0.$$

209. Suppose we have a function of two independent variables, and its partial derivatives with respect to them, and wish

to introduce, in place of our old variables, two others connnected with them by given relations.

For example: let z be a function of x and y, and let it be required to introduce, instead of x and y, u and v, which are connected with x and y by given equations. If the equations can be readily solved so as to express u in terms of x and y, and v in terms of x and y, we may proceed as follows: —

After the substitution, z is to be an explicit function of u and v. Suppose the substitution performed. As u and v are functions of x, z is indirectly a function of x. To get $D_x u$, we suppose y constant, so that x is for the time being the only independent variable, and we can get $D_x z$, by Art. 199, which gives us

$$D_x z = D_u z\, D_x u + D_v z\, D_x v$$

where all the derivatives are partial derivatives. In the same way,
$$D_y z = D_u z\, D_y u + D_v z\, D_y v.$$

$D_x u$, $D_x v$, $D_y u$, and $D_y v$ are found from the values of u and v mentioned above, and are generally functions of x and y, and $D_x z$ and $D_y z$ are at first obtained in terms of u, v, x, and y. x and y must be replaced by u and v by the aid of the given equations, and $D_x z$ and $D_y z$ are then in terms of u and v alone. By extending the process, we can get $D_x^2 z$, $D_x D_y z$, $D_y^2 z$, &c., in terms of u and v.

For example: introduce u and v in place of x and y in the equation $\qquad D_x^2 z = D_y^2 z.$

Given
$$\left. \begin{aligned} u &= x + y \\ v &= x - y \end{aligned} \right\}$$

$$D_x u = 1,$$
$$D_x v = 1,$$
$$D_y u = 1,$$
$$D_y v = -1,$$

$$D_x z = D_u z + D_v z,$$

$$D_y z = D_u z - D_v z,$$

$$D_x^2 z = D_u^2 z + 2 D_u D_v z + D_v^2 z,$$

$$D_y^2 z = D_u^2 z - 2 D_u D_v z + D_v^2 z,$$

$$D_u^2 z + 2 D_u D_v z + D_v^2 z = D_u^2 z - 2 D_u D_v z + D_v^2 z,$$

$$4 D_u D_v z = 0.$$

$$D_u D_v z = 0, \quad \text{the required equation.}$$

210. If it is more convenient to express x and y in terms of u and v at the start, we can proceed thus: z is explicitly a function of x and y, and if we regard v as constant for the time being, z is indirectly a function of the single variable u. Hence,

$$D_u z = D_x z D_u x + D_y z D_u y\,;$$

in like manner, $\quad D_v z = D_x z D_v x + D_y z D_v y.$

$D_u x$, $D_u y$, $D_v x$, and $D_v y$ are found in terms of u and v, and then by elimination between the equations, we get $D_x z$ and $D_y z$ in terms of u and v.

Examples.

(1) Given
$$\left. \begin{array}{l} x = r \cos \varphi \\ y = r \sin \varphi \end{array} \right\}$$

$$z = f(x,y),$$

find $D_x z$ and $D_y z$ in terms of r and φ.

Solution:
$$D_r x = \cos \varphi,$$

$$D_\varphi x = - r \sin \varphi,$$

$$D_r y = \sin \varphi,$$

$$D_\phi y = r\cos\varphi,$$

$$D_r z = D_x z \cos\varphi + D_y z \sin\varphi,$$

$$D_\phi z = -D_x z r \sin\varphi + D_y z r \cos\varphi.$$

Eliminate;

$$r\cos\varphi\, D_r z = r\cos^2\varphi\, D_x z + r\sin\varphi\cos\varphi\, D_y z,$$

$$\sin\varphi\, D_\phi z = -r\sin^2\varphi\, D_x z + r\sin\varphi\cos\varphi\, D_y z;$$

$$D_x z = \frac{1}{r}(r\cos\varphi\, D_r z - \sin\varphi\, D_\phi z)$$

$$D_y z = \frac{1}{r}(r\sin\varphi\, D_r z + \cos\varphi\, D_\phi z).$$

(2) Solve this same example by the method of Art. 209, using the relations
$$\left.\begin{array}{l} r^2 = x^2 + y^2 \\ \tan\varphi = \dfrac{y}{x} \end{array}\right\}$$

211. If it is not convenient to solve the given equations between x, y, u, and v, we can use the general method of either of the preceding articles, obtaining our $D_x u$, $D_x v$, $D_y u$, and $D_y v$, or our $D_u x$, $D_u y$, $D_v x$, $D_v y$, as follows: We have given

$$F_1(x,y,u,v) = 0 \text{ and } F_2(x,y,u,v) = 0.$$

Suppose y constant, then u and v will be functions of x; and, by Art. 200,
$$\left.\begin{array}{l} D_x F_1 + D_u F_1 D_x u + D_v F_1 D_x v = 0 \\ D_x F_2 + D_u F_2 D_x u + D_v F_2 D_x v = 0 \end{array}\right\}$$

From these equations we can obtain $D_x u$ and $D_x v$, and from two equations formed in the same way we can get $D_y u$ and $D_y v$; and a like process would give us $D_u x$, $D_u y$, $D_v x$, $D_v y$.

Examples.

(1) If V is a function of v, and
$$v^2 = x^2 + y^2,$$
show that
$$D_x^2 V + D_y^2 V = \frac{d^2 V}{dv^2} + \frac{1}{v}\frac{dV}{dv}.$$

(2) If V is a function of v, and
$$v^2 = x^2 + y^2 + z^2,$$
show that
$$D_x^2 V + D_y^2 V + D_z^2 V = \frac{d^2 V}{dv^2} + \frac{2 dV}{v dv}.$$

(3) If $\quad x = ae^\theta \cos\varphi \quad$ and $\quad y = ae^\theta \sin\varphi,$

show that $y^2 D_x^2 u - 2xy\, D_x D_y u + x^2 D_y^2 u = D_\varphi^2 u + D_\theta u.$

(4) Given
$$e^x + e^y = s,$$
$$e^{-y} + e^{-x} = t,$$
express $D_x^2 u + 2 D_x D_y u + D_y^2 u$ in terms of s and t.

Ans. $s^2 D_s^2 u - 2st D_s D_t u + t^2 D_t^2 u + s D_s u + t D_t u.$

CHAPTER XIV.

TANGENT LINES AND PLANES.

212. It is shown in Analytic Geometry of Three Dimensions, that any equation $F(x,y,z) = 0$ represents a surface, and that two such equations,

$$F_1(x,y,z) = 0,$$
$$F_2(x,y,z) = 0,$$

regarded as simultaneous equations, represent a curve in space, the intersection of the surfaces which the equations separately represent.

By eliminating z between these two equations, we can express y as an explicit function of x; and by eliminating y, we can express z in terms of x: consequently, the equations of any curve in space may be written in the form,

$$\left. \begin{array}{l} y = fx \\ z = Fx \end{array} \right\}$$

213. Let it be required to find the direction of the tangent line drawn at any given point (x_0, y_0, z_0) of the curve

$$\left. \begin{array}{l} y = fx \\ z = Fx \end{array} \right\}$$

Let $(x_0 + \Delta x, y_0 + \Delta y, z_0 + \Delta z)$ be any second point on the given curve. The equations of the line joining the two points are

$$\frac{x-x_0}{\varDelta x} = \frac{y-y_0}{\varDelta y} = \frac{z-z_0}{\varDelta z},$$

by Analytic Geometry; or

$$\left.\begin{aligned}\frac{y-y_0}{x-x_0} &= \frac{\varDelta y}{\varDelta x} \\ \frac{z-z_0}{x-x_0} &= \frac{\varDelta z}{\varDelta x}\end{aligned}\right\}$$

Let $\varDelta x$ approach zero, and the secant line approaches the required tangent as its limit, and this will have for its equations,

$$\left.\begin{aligned}\frac{y-y_0}{x-x_0} &= \left(\frac{dy}{dx}\right)_{x=x_0} \\ \frac{z-z_0}{x-x_0} &= \left(\frac{dz}{dx}\right)_{x=x_0}\end{aligned}\right\}$$

or, writing them in a more symmetrical form,

$$\frac{x-x_0}{1} = \frac{y-y_0}{\dfrac{dy_0}{dx_0}} = \frac{z-z_0}{\dfrac{dz_0}{dx_0}},$$

where, by $\dfrac{dy_0}{dx_0}$, we mean the value $\dfrac{dy}{dx}$ has when $x=x_0$.

A plane through the given point perpendicular to the tangent line is called the *normal* plane at the point in question. Prove that its equation is

$$x - x_0 + (y-y_0)\frac{dy_0}{dx_0} + (z-z_0)\frac{dz_0}{dx_0} = 0.$$

Example.

214. The helix is a curve traced on the surface of a cylinder of revolution by a point revolving about the axis of the cylinder

at a uniform rate, and at the same time advancing with a uniform velocity in the direction of the axis.

We can easily express its equations by the aid of an auxiliary

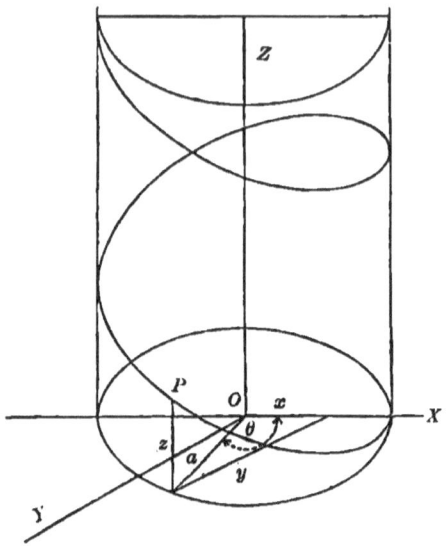

angle, the angle through which the point has rotated. Calling this angle θ and the radius of the circle a, we readily see that

$$x = a \cos \theta,$$

$$y = a \sin \theta.$$

From the nature of the helix, z must be proportional to the angle θ; hence $\dfrac{z}{\theta} = k,$ a constant,

and $z = k\theta.$

The required equations are then

$$\left.\begin{array}{l} x = a \cos \theta \\ y = a \sin \theta \\ z = k\theta \end{array}\right\}$$

CHAP. XIV.] TANGENT LINES AND PLANES.

To find the tangent line and normal plane at (x_0, y_0, z_0),

$$dy = a \cos \theta \, d\theta,$$

$$dx = - a \sin \theta \, d\theta,$$

$$\frac{dy}{dx} = - \ctn \theta = - \frac{x}{y},$$

$$dz = k \, d\theta,$$

$$\frac{dz}{dx} = - \frac{k}{a \sin \theta} = - \frac{k}{y}.$$

The equations of the tangent are

$$\frac{x - x_0}{1} = \frac{y - y_0}{-\dfrac{x_0}{y_0}} = \frac{z - z_0}{-\dfrac{k}{y_0}}$$

or

$$\frac{x - x_0}{- y_0} = \frac{y - y_0}{x_0} = \frac{z - z_0}{k}. \qquad [1]$$

The normal plane is

$$y_0 (x - x_0) - x_0 (y - y_0) - k (z - z_0) = 0. \qquad [2]$$

The direction cosines of line [1] are, by Analytic Geometry,

$$\cos \alpha = \frac{- y_0}{\sqrt{(x_0^2 + y_0^2 + k^2)}}$$

or

$$\cos \alpha = \frac{- y_0}{\sqrt{(a^2 + k^2)}},$$

$$\cos \beta = \frac{x_0}{\sqrt{(a^2 + k^2)}},$$

$$\cos \gamma = \frac{k}{\sqrt{(a^2 + k^2)}},$$

Cos γ is, then, not dependent on the position of the point P; therefore the helix has everywhere the same inclination to the axis of the cylinder; or, in other words, it crosses all the elements of the cylindrical surface at the same angle. If, then, this surface is unrolled into a plane surface, the helix will develop into a straight line.

215. The equations of the tangent line to the curve

$$f(x,y,z) = 0,$$

$$F(x,y,z) = 0,$$

can be obtained in a very convenient form if we use partial derivatives. We have, by Art. 199,

$$\left.\begin{aligned}\frac{df}{dx} &= D_x f + D_y f \frac{dy}{dx} + D_z f \frac{dz}{dx} = 0 \\ \frac{dF}{dx} &= D_x F + D_y F \frac{dy}{dx} + D_z F \frac{dz}{dx} = 0\end{aligned}\right\} \quad (1)$$

From these equations we can obtain the values of $\dfrac{dy}{dx}$ and $\dfrac{dz}{dx}$. Substituting these in the equations of Art. 213, and reducing, we get

$$\left.\begin{aligned}(x - x_0) D_{x_0} f + (y - y_0) D_{y_0} f + (z - z_0) D_{z_0} f = 0 \\ (x - x_0) D_{x_0} F + (y - y_0) D_{y_0} F + (z - z_0) D_{z_0} F = 0\end{aligned}\right\}$$

as the equations of the required tangent. The same result may be obtained much more easily by substituting in (1) the values of $\dfrac{dy}{dx}$ and $\dfrac{dz}{dx}$ given by the equations in Art. 213.

CHAP. XIV.] TANGENT LINES AND PLANES. 225

EXAMPLES.

(1) Given
$$\left.\begin{array}{l} x^2 + y^2 - ax = 0 \\ ax + z^2 - a^2 = 0 \end{array}\right\}$$

as the equations of a curve, find the tangent at (x_0, y_0, z_0).

$$\text{Ans. } \left.\begin{array}{l} x_0 x + y_0 y + z_0 z = a^2 \\ a(x - x_0) + 2(z - z_0) z_0 = 0 \end{array}\right\}$$

(2) Given the circle
$$\left.\begin{array}{l} x^2 + y^2 + z^2 = a^2 \\ x + z = a \end{array}\right\}$$

find the tangent at (x_0, y_0, z_0).

$$\text{Ans. } \left.\begin{array}{l} x_0 x + y_0 y + z_0 z = a^2 \\ x + z = a \end{array}\right\}$$

216. The osculating plane at a given point of a curve in space is the limiting position approached by a plane through the point and two other points of the curve as the latter approach indefinitely near the given point.

If (x_0, y_0, z_0) is the given point, and we regard x as our independent variable, we can represent two other points of the curve (Art. 190) by

$$(x_0 + \Delta x, y_0 + \Delta y, z_0 + \Delta z)$$

and $\quad (x_0 + 2\Delta x, y_0 + 2\Delta y + \Delta^2 y, z_0 + 2\Delta z + \Delta^2 z)$.

Forming the equation of the plane through these three points, dividing by Δx^2, and taking the limiting values as Δx approaches zero, we shall get as the osculating plane,

$$(x - x_0)\left(\frac{dy}{dx}\frac{d^2z}{dx^2} - \frac{dz}{dx}\frac{d^2y}{dx^2}\right) - (y - y_0)\left(\frac{d^2z}{dx^2}\right) + (z - z_0)\frac{d^2y}{dx^2} = 0.$$

EXAMPLE.

Obtain the osculating plane of the helix at (x_0, y_0, z_0).

217. The tangent plane at a given point of any surface

$$f(x,y,z) = 0$$

can be found by the aid of the equations of Art. 215.

Let
$$F(x,y,z) = 0$$

be any second surface whatever passing through (x_0, y_0, z_0).

The tangent line to the curve of intersection of the two surfaces at the point (x_0, y_0, z_0), that is, to any curve through (x_0, y_0, z_0) traced on the given surface, has for its equations

$$\left. \begin{array}{l} (x-x_0) D_{x_0} f + (y-y_0) D_{y_0} f + (z-z_0) D_{z_0} f = 0 \\ (x-x_0) D_{x_0} F + (y-y_0) D_{y_0} F + (z-z_0) D_{z_0} F = 0 \end{array} \right\}$$

It therefore lies in the plane represented by the first of these equations, which must then be the required tangent plane,

$$(x-x_0) D_{x_0} f + (y-y_0) D_{y_0} f + (z-z_0) D_{z_0} f = 0.$$

Examples.

(1) Find the tangent plane to a sphere.

$$x^2 + y^2 + z^2 = a^2.$$

Ans. $x_0 x + y_0 y + z_0 z = a^2.$

(2) Find the tangent plane to an ellipsoid.

$$\frac{x^2}{a^2} + \frac{y^2}{b^2} + \frac{z^2}{c^2} = 1.$$

Ans. $\dfrac{x_0 x}{a^2} + \dfrac{y_0 y}{b^2} + \dfrac{z_0 z}{c^2} = 1.$

The normal line at (x_0, y_0, z_0) is easily seen to be

$$\frac{x - x_0}{D_{x_0} f} = \frac{y - y_0}{D_{y_0} f} = \frac{z - z_0}{D_{z_0} f}.$$

CHAPTER XV.

DEVELOPMENT OF A FUNCTION OF SEVERAL VARIABLES.

218. To develop $f(x+h, y+k)$ into a series arranged according to the powers of h and k, where h and k are any arbitrary increments that may be given. Let a be any variable, and call

$$\frac{h}{a} = h_1, \quad \frac{k}{a} = k_1,$$

so that $\qquad h = ah_1$ and $k = ak_1.$

If now x and y are regarded as given values, $f(x+h, y+k)$ is a function of h and k, which depend on a; and hence $f(x+h, y+k)$ can be considered a function of a. Call it Fa, and it may be developed by Maclaurin's Theorem, which gives

$$Fa = F0 + aF'0 + \frac{a^2}{2!} F''0 + \frac{a^3}{3!} F'''0 + \ldots$$

$$+ \frac{a^n}{n!} F^{(n)}0 + \frac{a^{n+1}}{(n+1)!} F^{(n+1)}\theta a.$$

When $\qquad a = 0,$

$$Fa \text{ or } F(x+h, y+k) = f(x, y).$$

Call $\qquad x + ah_1 = x'$ and $y + ak_1 = y',$

then $\qquad Fa = f(x', y'),$

$$F'a = \frac{df(x', y')}{da} = D_{x'} f(x', y') \frac{dx'}{da} + D_{y'} f(x', y') \frac{dy'}{da},$$

by Art. 199;

$$\frac{dx'}{da} = h_1,$$

$$\frac{dy'}{da} = k_1,$$

$$F'a = h_1 D_{x'} f(x', y') + k_1 D_{y'} f(x', y'),$$

$$F'0 = h_1 D_x f(x, y) + k_1 D_y f(x, y),$$

which we shall write $h_1 D_x f + k_1 D_y f$.

$$F''a = \frac{dF'a}{da} = h_1 D_{x'} F'a + k_1 D_{y'} F'a$$

$$= h_1^2 D_{x'}^2 f(x', y') + 2 h_1 k_1 D_{x'} D_{y'} f(x', y') + k_1^2 D_{y'}^2 f(x', y'),$$

$$F'''a = h_1^3 D_{x'}^3 f(x', y') + 3 h_1^2 k_1 D_{x'}^2 D_{y'} f(x', y')$$

$$+ 3 h_1 k_1^2 D_{x'} D_{y'}^2 f(x', y') + k_1^3 D_{y'}^3 f(x', y').$$

In $F''a$ and $F'''a$ the terms have a striking resemblance to the terms of the second and third powers of a binomial. Let us see whether this will hold for higher derivatives. Assume that it holds for the $F^{(n)}a$, and see if it holds for the $F^{(n+1)}a$.

If $$F^{(n)}a = h_1^n D_{x'}^n f(x', y') + n h_1^{n-1} k_1 D_{x'}^{n-1} D_{y'} f(x', y')$$

$$+ \frac{n(n-1)}{2!} h_1^{n-2} k_1^2 D_{x'}^{n-2} D_{y'}^2 f(x', y')$$

$$+ \frac{n(n-1)(n-2)}{3!} h_1^{n-3} k_1^3 D_{x'}^{n-3} D_{y'}^3 f(x', y') + \cdots$$

$$F^{(n+1)}a = h_1 D_{x'} F^{(n)}a + k_1 D_{y'} F^{(n)}a$$

$$= h_1^{n+1} D_{x'}^{n+1} f(x', y') + (n+1) h_1^n k_1 D_{x'}^n D_{y'} f(x', y')$$

$$+ \frac{(n+1)n}{2!} h_1^{n-1} k_1^2 D_{x'}^{n-1} D_{y'}^2 f(x', y') + \cdots$$

CHAP. XV.] FUNCTION OF SEVERAL VARIABLES. 229

If, then, the observed analogy holds for any derivative, it holds for the next higher. It does hold for the third; it holds then for the fourth, and for all succeeding ones.

$$F''0 = h_1^2 D_x^2 f + 2h_1 k_1 D_x D_y f + k_1^2 D_y^2 f,$$

$$F^{(n)} 0 = h_1^n D_x^n f + n h_1^{n-1} k_1 D_x^{n-1} D_y f$$

$$+ \frac{n(n-1)}{2!} h_1^{n-2} k_1^2 D_x^{n-2} D_y^2 f + \cdots$$

$$F^{(n+1)} \theta a = h_1^{n+1} D_x^{n+1} f(x + \theta h, y + \theta k)$$

$$+ (n+1) h_1^n k_1 D_x^n D_y f(x + \theta h, y + \theta k) + \cdots$$

By this notation we mean that $x + \theta h$, $y + \theta k$, are to be substituted for x and y after the differentiations are performed.

We have then, remembering that

$$a h_1 = h \text{ and } a k_1 = k,$$

$$f(x + h, y + k) = f(x, y) + (h D_x f + k D_y f)$$

$$+ \frac{1}{2!} (h^2 D_x^2 f + 2 h k D_x D_y f + k^2 D_y^2 f)$$

$$+ \frac{1}{3!} (h^3 D_x^3 f + 3 h^2 k D_x^2 D_y f + 3 h k^2 D_x D_y^2 f + k^3 D_y^3 f) + \cdots$$

$$+ \frac{1}{n!} \left(h^n D_x^n f + n h^{n-1} k D_x^{n-1} D_y f + \frac{n(n-1)}{2!} h^{n-2} k^2 D_x^{n-2} D_y^2 f + \cdots \right)$$

$$+ \frac{1}{(n+1)!} \left(h^{n+1} D_x^{n+1} f(x + \theta h, y + \theta k) \right.$$

$$+ (n+1) h^n k D_x^n D_y f(x + \theta h, y + \theta k)$$

$$\left. + \frac{(n+1)n}{2} h^{n-1} k^2 D_x^{n-1} D_y^2 f(x + \theta h, y + \theta k) + \cdots \right).$$

If we use $(h D_x + k D_y)^n f$ as an abbreviation for $(h^n D_x^n f + n h^{n-1} k D_x^{n-1} D_y f + \cdots)$; that is, understanding that $(h D_x + k D_y)^n$

is to be expanded just as though it were a binomial, and then to have each term written before f, we can simplify the above expression.

$$f(x+h, y+k) = f(x,y) + (hD_x + kD_y)f(x,y)$$

$$+ \frac{1}{2!}(hD_x + kD_y)^2 f(x,y) + \frac{1}{3!}(hD_x + kD_y)^3 f(x,y) + \cdots$$

$$+ \frac{1}{n!}(hD_x + kD_y)^n f(x,y)$$

$$+ \frac{1}{(n+1)!}(hD_x + kD_y)^{n+1} f(x+\theta h, y+\theta k),$$

which is Taylor's Theorem for two independent variables.

If we let $x = 0$ and $y = 0$,

we get $\quad f(h,k) = f(0,0) + (hD_x + kD_y)f(0,0)$

$$+ \frac{1}{2!}(hD_x + kD_y)^2 f(0,0) + \cdots;$$

or, changing h and k to x and y,

$$f(x,y) = f(0,0) + (xD_x + yD_y)f(0,0)$$

$$+ \frac{1}{2!}(xD_x + yD_y)^2 f(0,0) + \frac{1}{3!}(xD_x + yD_y)^3 f(0,0) + \cdots$$

$$+ \frac{1}{n!}(xD_x + yD_y)^n f(0,0) + \frac{1}{(n+1)!}(xD_x + yD_y)^{n+1} f(\theta x, \theta y),$$

which is Maclaurin's Theorem for two variables.

EXAMPLE.

Transform $Ax^2 + Bxy + Cy^2 + Dx + Ey + F = 0$

to (x_0, y_0) as a new origin, the formulas for transformation being

$$x = x_0 + x', \; y = y_0 + y'.$$

Call our given equation $f(x,y)$; we wish to develop

$$f(x_0+x', y_0+y').$$

$$f(x_0+x', y_0+y') = f(x_0,y_0) + (x'D_{x_0} + y'D_{y_0})f(x_0,y_0)$$

$$+ \frac{1}{2!}(x'D_{x_0} + y'D_{y_0})^2 f(x_0,y_0) + \cdots$$

$$D_x f(x,y) = 2Ax + By + D,$$

$$D_y f(x,y) = Bx + 2Cy + E,$$

$$D_x^2 f(x,y) = 2A,$$

$$D_x D_y f(x,y) = B,$$

$$D_y^2 f(x,y) = 2C;$$

all higher derivatives are 0.

$$f(x_0+x', y_0+y') = Ax_0^2 + Bx_0 y_0 + Cy_0^2 + Dx_0 + Ey_0 + F$$

$$+ (2Ax_0 + By_0 + D)x' + (Bx_0 + 2Cy_0 + E)y'$$

$$+ Ax'^2 + Bx'y' + Cy'^2, \qquad \text{a familiar result.}$$

219. By like reasoning, Taylor's Theorem can be extended to functions of more than two variables. For three variables it becomes

$$f(x+h, y+k, z+l) = f(x,y,z) + (hD_x + kD_y + lD_z)f(x,y,z)$$

$$+ \frac{1}{2!}(hD_x + kD_y + lD_z)^2 f(x,y,z)$$

$$+ \frac{1}{3!}(hD_x + kD_y + lD_z)^3 f(x,y,z) + \cdots$$

$$+ \frac{1}{(n+1)!}(hD_x + kD_y + lD_z)^{n+1} f(x+\theta h, y+\theta k, z+\theta l).$$

EXAMPLE.

Transform $x^2 + y^2 + z^2 - 4x + 6y - 2z - 11 = 0$

to the new origin $(2, -3, 1)$. Ans. $x^2 + y^2 + z^2 = 25$.

Euler's Theorem for Homogeneous Functions.

220. A homogeneous function of several variables is one of such a nature that, if each variable be multiplied by some constant, the function is multiplied by a power of that constant. The order of the function is the power of the constant by which it is multiplied.

For example: $x^2 + xy - y^2$ is homogeneous of the second order; for, if we change x into kx and y into ky, our function becomes $k^2(x^2 + xy - y^2)$, and is multiplied by the second power of k. $\sin\dfrac{x-y}{2x}$ is homogeneous of the zero order; for, if we multiply x and y by k, the function is unchanged; that is, it is multiplied by k^0.

Let $f(x,y)$ be a homogeneous function of x and y; then, no matter what the value of q,

$$f(x + qx, y + qy) = f(x,y) + q(xD_x + yD_y)f(x,y)$$
$$+ \frac{q^2}{2!}(xD_x + yD_y)^2 f(x,y) + \cdots$$
$$+ \frac{q^{m+1}}{(m+1)!}(xD_x + yD_y)^{m+1} f(x + q\theta x, y + q\theta y);$$

but $f(x + qx, y + qy) = f[(1+q)x, (1+q)y] = (1+q)^n f(x,y)$

by the definition of a homogeneous function.

Call $f(x,y) = u$, and we have

$$(1+q)^n u = u + q(xD_x + yD_y)u + \frac{q^2}{2!}(xD_x + yD_y)^2 u$$
$$+ \frac{q^3}{3!}(xD_x + yD_y)^3 u + \cdots.$$

As this equation must hold, no matter what the value of q, the coefficients of like powers of q in the two members of the equation must be the same. Equating them, we have

$$u = u,$$

$$(xD_x + yD_y)u = nu,$$

$$(xD_x + yD_y)^2 u = n(n-1)u,$$

$$(xD_x + yD_y)^3 u = n(n-1)(n-2)u,$$

$$(xD_x + yD_y)^m u = n(n-1)(n-2)\cdots(n-m+1)u;$$

and these equations are Euler's Theorem.

Examples.

Verify Euler's Theorem for second and third derivatives when

$$u = x^3 + y^3 \text{ and when } u = \sin^{-1}\frac{y}{x}.$$

CHAPTER XVI.

MAXIMA AND MINIMA OF FUNCTIONS OF TWO OR MORE VARIABLES.

221. If we have a function of two variables $u = f(x,y)$, and $f(x_0 + h, y_0 + k) - f(x_0, y_0) < 0$ for small values of h and k, no matter what the signs and relative magnitudes of these values, u is a *maximum* for the values x_0, y_0 of x and y. If $f(x_0 + h, y_0 + k) - f(x_0 y_0) > 0$ under these same circumstances, u is a *minimum*. By Taylor's Theorem,

$$f(x_0 + h, y_0 + k) - f(x_0, y_0) = (hD_x + kD_y) f(x_0, y_0)$$

$$+ \frac{1}{2!} (hD_x + kD_y)^2 f(x_0 + \theta h, y_0 + \theta k).$$

If we take the values of h and k sufficiently small, we can always make $\frac{1}{2}(hD_x + kD_y)^2 f(x_0 + \theta h, y_0 + \theta k) < (hD_x + kD_y) f(x_0, y_0)$, and then the sign of the second member will be the sign of $(hD_x + kD_y) f(x_0, y_0)$; that is, of $hD_{x_0} u_0 + kD_{y_0} u_0$, which evidently depends upon the signs of h and k. In order, then, that the sign of $f(x_0 + h, y_0 + k) - f(x_0, y_0)$ should be constant, — that is, in order that for x_0, y_0 u should be either a maximum or a minimum, — the terms $hD_x u + kD_y u$ must disappear, no matter what the values of h and k; or, in other words, $D_{x_0} u_0$ and $D_{y_0} u_0$ must both equal 0. We get, then, as essential to the existence of either a maximum or a minimum, the conditions

$$D_x u = 0,$$

$$D_y u = 0,$$

for the values of x and y in question.

CHAP. XVI.] MAXIMA AND MINIMA OF FUNCTIONS. 235

222. Carrying the development a step farther, and assuming that $D_{x_0} u_0$ and $D_{y_0} u_0$ are zero,

$$f(x_0+h, y_0+k) - f(x_0, y_0) = \frac{1}{2!}(hD_x + kD_y)^2 f(x_0, y_0)$$

$$+ \frac{1}{3!}(hD_x + kD_y)^3 f(x_0 + \theta h, y_0 + \theta k).$$

As before, it is evident that for small values of h and k, the sign of the whole second member will be that of the terms $\frac{1}{2}(h^2 D_{x_0}^2 u_0 + 2hk D_{x_0} D_{y_0} u_0 + k^2 D_{y_0}^2 u_0)$. Let us investigate this carefully.

Let
$$A = D_{x_0}^2 u_0,$$
$$B = D_{x_0} D_{y_0} u_0,$$
$$C = D_{y_0}^2 u_0,$$

our parenthesis becomes $Ah^2 + 2Bhk + Ck^2$; and for a maximum or minimum the sign of this must be independent of the signs and values of h and k.

$$Ah^2 + 2Bhk + Ck^2 = \frac{1}{A}(A^2 h^2 + 2ABhk + ACk^2),$$

$$= \frac{1}{A}(A^2 h^2 + 2ABhk + B^2 k^2 - B^2 k^2 + ACk^2),$$

$$= \frac{1}{A}[(Ah + Bk)^2 + (AC - B^2)k^2].$$

$(Ah + Bk)^2$ and k^2 are necessarily positive. If $AC - B^2$ is also positive, the sign of the whole expression will be independent of h and k, and will be positive if A is positive, and negative if A is negative. If $\quad AC - B^2 = 0,\quad$ the result is the same;

but if $AC - B^2$ is negative, the sign of the parenthesis will depend upon the sign and relative values of h and k, and we shall have neither a maximum nor a minimum.

223. To sum up: —

If
$$D_{x_0} u_0 = 0$$
$$D_{y_0} u_0 = 0$$
$$D_{x_0}^2 u_0 D_{y_0}^2 u_0 - (D_{x_0} D_{y_0} u_0)^2 = \text{ or } > 0$$
$$D_{x_0}^2 u_0 < 0$$
$\Big\}$ u_0 is a maximum.

If
$$D_{x_0} u_0 = 0$$
$$D_{y_0} u_0 = 0$$
$$D_{x_0}^2 u_0 D_{y_0}^2 u_0 - (D_{x_0} D_{y_0} u_0)^2 = \text{ or } > 0$$
$$D_{x_0}^2 u_0 > 0$$
$\Big\}$ u_0 is a minimum.

Examples.

224. (1) To find a point so situated that the sum of the squares of its distances from the three vertices of a given triangle shall be a minimum. Let (x_1, y_1), (x_2, y_2), (x_3, y_3) be the given vertices, and (x, y) the required point.

$$u = (x - x_1)^2 + (y - y_1)^2 + (x - x_2)^2 + (y - y_2)^2$$
$$+ (x - x_3)^2 + (y - y_3)^2$$

is the function which we must make a minimum.

$$D_x u = 2(x - x_1) + 2(x - x_2) + 2(x - x_3),$$
$$D_y u = 2(y - y_1) + 2(y - y_2) + 2(y - y_3),$$
$$D_x^2 u = 2 + 2 + 2 = 6 = A,$$
$$D_x D_y u = 0 = B,$$
$$D_y^2 u = 2 + 2 + 2 = 6 = C.$$

We must make $D_x u$ and $D_y u$ both equal to zero.

$$2(x-x_1)+2(x-x_2)+2(x-x_3)=0,$$

$$x=\frac{x_1+x_2+x_3}{3},$$

$$2(y-y_1)+2(y-y_2)+2(y-y_3)=0,$$

$$y=\frac{y_1+y_2+y_3}{3},$$

$$AC-B^2=36-0>0,$$

$$A=6>0.$$

Hence u is a minimum when

$$x=\frac{x_1+x_2+x_3}{3} \text{ and } y=\frac{y_1+y_2+y_3}{3}.$$

The required point is the centre of gravity of the triangle.

(2) To inscribe in a circle a triangle of maximum perimeter. Join the centre with each vertex and with the middle point of

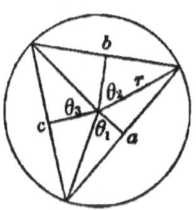

each side. The angles between the three radii are bisected by the lines drawn to the middle points of the sides. Call these half-angles $\theta_1, \theta_2, \theta_3$.

$$\frac{a}{2} \div r = \sin\theta_1,$$

$$a=2r\sin\theta_1,$$

$$b=2r\sin\theta_2,$$

$$c=2r\sin\theta_3,$$

$$2\theta_1 + 2\theta_2 + 2\theta_3 = 2\pi,$$

$$\theta_1 + \theta_2 + \theta_3 = \pi, \tag{1}$$

$$p = a + b + c = 2r(\sin\theta_1 + \sin\theta_2 + \sin\theta_3)$$

is the function we are to make a maximum, and is a function of two independent variables, say θ_1 and θ_2; for we can regard θ_3 as depending on θ_1 and θ_2 through equation (1). As r is a constant, it will be enough to make

$$u = \sin\theta_1 + \sin\theta_2 + \sin\theta_3 \qquad \text{a maximum.}$$

$$D_{\theta_1} u = \cos\theta_1 + \cos\theta_3 D_{\theta_1}\theta_3;$$

for, since $\qquad \theta_1 + \theta_2 + \theta_3 = \pi,$

changing θ_1 without changing θ_2 will change θ_3.

$$D_{\theta_1}\theta_3 = -1;$$

hence $\qquad D_{\theta_1} u = \cos\theta_1 - \cos\theta_3.$

$$D_{\theta_2} u = \cos\theta_2 - \cos\theta_3,$$

for $\qquad D_{\theta_2}\theta_3 = -1,$

$$D_{\theta_1}^2 u = -\sin\theta_1 - \sin\theta_3,$$

$$D_{\theta_1} D_{\theta_2} u = -\sin\theta_3,$$

$$D_{\theta_2}^2 u = -\sin\theta_2 - \sin\theta_3.$$

Make $\qquad D_{\theta_1} u = 0$ and $D_{\theta_2} u = 0.$

$$\left.\begin{array}{c} \cos\theta_1 - \cos\theta_3 = 0 \\ \cos\theta_2 - \cos\theta_3 = 0 \end{array}\right\}$$

$$\theta_1 = \theta_2 = \theta_3.$$

Substitute these values in $D_{\theta_1}^2 u$, &c., and

$$D_{\theta_1}^2 u = -2\sin\theta_1 = A,$$

$$D_{\theta_1} D_{\theta_2} u = +\sin\theta_1 = B,$$

$$D_{\theta_2}^2 u = -2\sin\theta_1 = C,$$

$$AC - B^2 = 4\sin^2\theta_1 - \sin^2\theta_1 = 3\sin^2\theta_1 > 0,$$

$$A = -2\sin\theta_1 < 0, \text{ and } u \text{ is a maximum.}$$

Since
$$\theta_1 = \theta_2 = \theta_3,$$

$$a = b = c;$$

and the required triangle is equilateral.

(3) To inscribe in a circle a triangle of maximum area.
Ans. The triangle is equilateral.

225. Very often it is unnecessary to examine the second derivatives, as the nature of the problem enables one to determine whether the value of the variables obtained by writing the first derivatives equal to zero corresponds to maximum or minimum values of the function.

EXAMPLES.

(1) Required the form of a parallelopiped of given volume and minimum surface. *Ans.* A cube.

(2) Required the form of a parallelopiped of given surface and maximum volume. *Ans.* A cube.

(3) An open cistern in the form of a parallelopiped is to be built, capable of containing a given volume of water, what must be its form that the expense of lining its interior surface may be a minimum?
Ans. Length and breadth each double the depth.

CHAPTER XVII.

THEORY OF PLANE CURVES.

Concavity and Convexity.

226. The words concavity and convexity are used in mathematics in their ordinary sense. A curve is concave toward the axis of X when it bends toward the axis; convex, when it bends from the axis: that is, when in passing along the curve its inclination to the axis of X decreases, the curve is concave; when it increases, the curve is convex, supposing that the portion of the curve considered lies above the axis; if it lies below the axis, the rule just given must be reversed. We have seen that the tangent of this inclination, which we have called τ, is equal to $\frac{dy}{dx}$. If the curve is concave and above the axis, τ decreases as we increase x, $\tan \tau$ or $\frac{dy}{dx}$ decreases, and $\frac{d^2y}{dx^2} < 0$, by Art. 37. If the curve is convex, $\frac{d^2y}{dx^2} > 0$.

227. A point at which the curve is changing from convexity to concavity, or from concavity to convexity, is called a *point of inflection*. At such a point, $\frac{d^2y}{dx^2}$ is changing from a negative to a positive value, or from a positive to a negative value, and consequently must be passing through the value zero. To sum up: if
$$y = fx$$
is the equation of a plane curve, at any point corresponding to

CHAP. XVII.]   THEORY OF PLANE CURVES.   241

a value of x that makes $\frac{d^2y}{dx^2} < 0$, the curve is concave towards the axis of x, if above the axis; convex, if below. At any point corresponding to a value of x that makes $\frac{d^2y}{dx^2} > 0$, the curve is convex towards the axis of x, if above the axis; concave, if below. Any point corresponding to a value of x that makes

$$\frac{d^2y}{dx^2} = 0$$

is in general a point of inflection.

We have seen that the curvature,

$$k = \frac{-\frac{d^2y}{dx^2}}{\left[1 + \left(\frac{dy}{dx}\right)^2\right]^{\frac{3}{2}}}.$$

It is easily seen that at a point of inflection this value changes sign.

228. These same tests for concavity, convexity, and inflection can be very simply obtained by the aid of Taylor's Theorem.

Let
$$y = fx$$

be the equation of a curve, and let it be required to discover whether the curve is concave or convex toward the axis of X at the point corresponding to the value

$$x = a.$$

Draw a tangent at the point in question, and erect ordinates to the curve and to the tangent near the point of contact.

It is evident that the ordinate of a point in the curve minus the ordinate of the corresponding point of the tangent must be negative on both sides of the point of contact, if the curve is *concave*, and positive on both sides of the point of contact, if the curve is *convex*. If the point is a *point of inflection*, this difference will have opposite signs on different sides of the point.

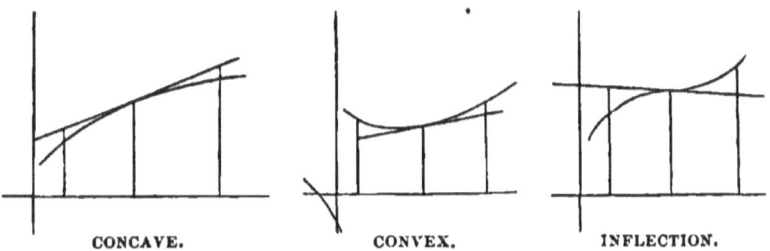

CONCAVE.   CONVEX.   INFLECTION.

The equation of the tangent at the point corresponding to

$$x = a$$

is $\qquad y - fa = f'a(x - a), \qquad$ by Art. 28, [1].

Let $\qquad x = a + h$

in the equations of curve and tangent, and call the corresponding values of y, y_1 and y_2; then

$$y_2 = fa + hf'a.$$

$$y_1 = fa + hf'a + \frac{h^2}{2!} f''a + \frac{h^3}{3!} f'''(a + \theta h),$$

by Taylor's Theorem.

$$y_1 - y_2 = \frac{h^2}{2!} f''a + \frac{h^3}{3!} f'''(a + \theta h).$$

If $f''a$ does not equal zero, h may be taken so small that the sign of $y_1 - y_2$ will be the sign of $\frac{h^2}{2!} f''a$.

If $f''a$ is positive this sign is positive whether h is positive or negative, and the curve is convex. If $f''a$ is negative, $y_1 - y_2$ is negative both before and after $x = a$, and the curve is concave.

If $f''a = 0$ and $f'''a$ does not vanish, the sign of $y_1 - y_2$ will change as the sign of h changes, and we shall have a point of inflection.

CHAP. XVII.] THEORY OF PLANE CURVES. 243

229. For example, let us see whether the curve

$$x^2 + y^2 = 25$$

is convex or concave towards the axis of X at the point $(3,4)$.

$$2x\,dx + 2y\,dy = 0. \tag{1}$$

$$2dx^2 + 2dy^2 + 2y\,d^2y = 0. \tag{2}$$

From (1)
$$dy = -\frac{x\,dx}{y}.$$

Substitute in (2), $2dx^2 + \dfrac{2x^2\,dx^2}{y^2} + 2y\,d^2y = 0,$

$$(x^2 + y^2)dx^2 + y^3\,d^2y = 0,$$

$$25\,dx^2 + y^3\,d^2y = 0.$$

$$\frac{d^2y}{dx^2} = -\frac{25}{y^3} = -\frac{25}{64} < 0$$

at the point $(3,4)$; and the curve is concave.

Again, let us see whether the curve

$$y = x(x-a)^4 \quad \text{has points of inflection}$$

$$\frac{dy}{dx} = (x-a)^4 + 4x(x-a)^3,$$

$$\frac{d^2y}{dx^2} = 8(x-a)^3 + 12x(x-a)^2,$$

$$\frac{d^3y}{dx^3} = 36(x-a)^2 + 24x(x-a),$$

$$\frac{d^4y}{dx^4} = 96(x-a) + 24x.$$

Write
$$\frac{d^2y}{dx^2} = 0.$$

and we get $\quad 8(x-a)^3 + 12x(x-a)^2 = 0$

or $\quad 2(x-a)^3 + 3x(x-a)^2 = 0.$

One root is $\quad x = a;$
divide by $(x-a)^2$, and
$$2x - 2a + 3x = 0.$$

$$x = \frac{2a}{5} \quad \text{is the remaining root.}$$

If $\quad x = \dfrac{2a}{5},$

$\dfrac{d^3y}{dx^3}$ does not equal zero, and we get a point of inflection.

If $\quad x = a,$

$$\frac{d^3y}{dx^3} = 0,$$

$\dfrac{d^4y}{dx^4}$ does not equal zero, and the point is not a point of inflection.

EXAMPLES.

(1) If $\quad y = \dfrac{x^3}{a^2 + x^2},$

there is a point of inflection at the origin, and also when $x = \pm a\sqrt{(3)}$.

(2) If $\quad \dfrac{y}{a} = \sqrt{\left(\dfrac{a-x}{x}\right)},$

there is a point of inflection when $x = \dfrac{3a}{4}$.

(3) If $\quad x^{\frac{1}{2}} = \log y,$

there is a point of inflection when $x = 8$.

(4) If $$xy = a^2 \log \frac{x}{a},$$
there is a point of inflection when $x = ae^{\frac{1}{2}}$.

Singular Points.

230. Singular points of a curve are points possessing some peculiarity independent of the position of the axes. Such points are, —
1. Points of inflection (Art. 228);
2. Multiple points;
3. Cusps;
4. Conjugate points;
5. Points d'arrêt;
6. Points saillant.

231. (2) A *multiple point* is one through which two or more branches of the curve pass. If only two branches pass through

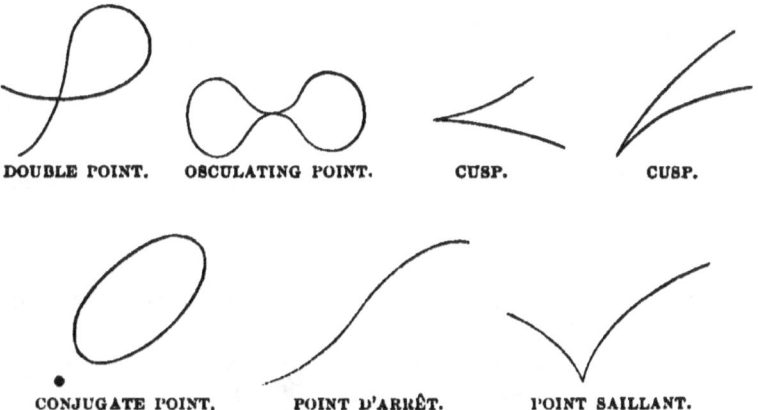

DOUBLE POINT.   OSCULATING POINT.   CUSP.   CUSP.

CONJUGATE POINT.   POINT D'ARRÊT.   POINT SAILLANT.

the point, it is a *double* point. A double point at which the branches of the curve are tangent is an *osculating* point.

(3) An osculating point where both branches of the curve stop is a *cusp*.

(4) An isolated point of a curve is a *conjugate point*.

(5) A point at which a single branch of a curve suddenly stops is a *point d'arrêt*.

(6) A double point at which the two branches of the curve stop without being tangent to each other is a *point saillant*.

Multiple Points.

232. At a multiple point, the curve will in general have more than one tangent, and therefore $\dfrac{dy}{dx}$ will have more than one value.

Let
$$\varphi = 0$$

be the equation of the curve in rational algebraic form.

$$\frac{dy}{dx} = \frac{-D_x\varphi}{D_y\varphi},$$

by Art. 202. For any given values of x and y, $D_x\varphi$ and $D_y\varphi$ will have each a definite value, as they are rational polynomials; and $\dfrac{dy}{dx}$ will have but one value, unless $D_x\varphi$ and $D_y\varphi$ are both zero, in which case $\dfrac{dy}{dx} = \dfrac{0}{0},$ and is indeterminate;

hence, our fundamental condition for the existence of a multiple point is
$$D_x\varphi = 0 \text{ and } D_y\varphi = 0.$$

To determine $\dfrac{dy}{dx}$ in that case, we differentiate numerator and denominator,
$$\frac{dy}{dx} = -\frac{D_x^2\varphi + D_x D_y\varphi \dfrac{dy}{dx}}{D_x D_y\varphi + D_y^2\varphi \dfrac{dy}{dx}}. \tag{1}$$

Clearing of fractions gives us
$$D_y^2\varphi \left(\frac{dy}{dx}\right)^2 + 2D_x D_y\varphi \frac{dy}{dx} + D_x^2\varphi = 0, \tag{2}$$

CHAP. XVII.] THEORY OF PLANE CURVES. 247

a quadratic to determine $\frac{dy}{dx}$. Unless (1) is still indeterminate, that is, unless $D_x^2\varphi$, $D_x D_y\varphi$, and $D_y^2\varphi$ are all zero, we get two values of $\frac{dy}{dx}$, and the point is a *double point*.

If $\frac{dy}{dx}$ is still indeterminate, we differentiate (1) again, and get

$$D_y^3\varphi\left(\frac{dy}{dx}\right)^3 + 3D_x D_y^2\varphi\left(\frac{dy}{dx}\right)^2 + 3D_x^2 D_y\varphi\frac{dy}{dx} + D_x^3\varphi = 0,$$

to determine $\frac{dy}{dx}$. We have then three tangents at the point, which is a *triple point*.

233. If the values of $\frac{dy}{dx}$ obtained from Art. 232 (2) are equal, the two tangents at the double point coincide, and the point is an *osculating point* or a *cusp*; and we cannot tell which except by actually tracing the curve in the neighborhood of the point.

If the two values of $\frac{dy}{dx}$ are imaginary, no tangent can be drawn at the point, which is then a *conjugate* point.

A *point d'arrêt* or a *point saillant* can be discovered only by inspection when attempting to trace the curve; they occur only in transcendental curves.

Example.

234. To investigate the existence of multiple points in the curve

$$x^4 - a^2x^2 + a^2y^2 = 0.$$

$$D_x\varphi = 4x^3 - 2a^2x,$$

$$D_y\varphi = 2a^2y,$$

$$D_x^2\varphi = 12x^2 - 2a^2.$$

$$D_x D_y \varphi = 0,$$

$$D_y^2 \varphi = 2a^2,$$

$D_x \varphi$ and $D_y \varphi$ must equal zero.

$$4x^3 - 2a^2 x = 0$$

if $\quad x = 0$ or if $x = \pm \dfrac{a}{\sqrt{(2)}}.$

$$2a^2 y = 0 \text{ if } y = 0;$$

hence x must equal zero, and y equal 0,

or $\quad x = \pm \dfrac{a}{\sqrt{(2)}}$ and $y = 0;$

but $\left(\pm \dfrac{a}{\sqrt{(2)}}, 0 \right)$ is not a point of the curve; therefore we need consider only $(0,0)$. In this case,

$$D_x^2 \varphi = -2a^2,$$

$$D_y^2 \varphi = 2a^2,$$

$$2a^2 \left(\frac{dy}{dx}\right)^2 - 2a^2 = 0,$$

$$\left(\frac{dy}{dx}\right)^2 = 1,$$

$$\frac{dy}{dx} = \pm 1,$$

and the origin is a double point of the curve, the two branches making with the axis of X angles of $45°$ and $135°$ respectively.

EXAMPLE.

235. Consider $x^3 - y^2 = 0$.

$$D_x \varphi = 3x^2,$$
$$D_y \varphi = -2y,$$
$$D_x^2 \varphi = 6x,$$
$$D_x D_y \varphi = 0,$$
$$D_y^2 \varphi = -2.$$

Make $3x^2 = 0$ and $-2y = 0$.

We get $\left. \begin{array}{l} x = 0 \\ y = 0 \end{array} \right\}$ as the only point we need consider here.

In this case,
$$D_x^2 \varphi = 0,$$
$$D_x D_y \varphi = 0,$$
$$D_y^2 \varphi = -2.$$
$$-2\left(\frac{dy}{dx}\right)^2 = 0,$$
$$\frac{dy}{dx} = \pm 0.$$

The values of $\frac{dy}{dx}$ are equal, and the origin is an *osculating point*, both branches being there tangent to the axis of X.

Since $y^2 = x^3$,

it is easily seen that the curve lies to the right, and not to the left, of the origin, which is therefore a *cusp*.

Example.

236.
$$bx^2 - x^3 + ay^2 = 0.$$
$$D_x \varphi = 2bx - 3x^2,$$
$$D_y \varphi = 2ay,$$
$$D_x^2 \varphi = 2b - 6x,$$
$$D_x D_y \varphi = 0,$$
$$D_y^2 \varphi = 2a,$$
$$\left.\begin{array}{l} 2bx - 3x^2 = 0 \\ 2ay = 0 \end{array}\right\}$$
$$\left.\begin{array}{l} x = 0 \\ y = 0 \end{array}\right\} \text{ is to be considered.}$$

At this point,
$$D_x^2 \varphi = 2b,$$
$$D_x D_y \varphi = 0,$$
$$D_y^2 \varphi = 2a,$$
$$2a\left(\frac{dy}{dx}\right)^2 + 2b = 0,$$
$$\left(\frac{dy}{dx}\right)^2 = \frac{-b}{a}.$$

If b and a have the same sign, $\frac{dy}{dx}$ is imaginary, and the origin is a *conjugate point*; a result that can be easily verified by examining the equation.

Examples.

(1) Show that the curve $y = x \log x$ has a point d'arrêt at the origin.

(2) Show that the curve $y = \dfrac{x}{1+e^{\frac{1}{x}}}$ has a point saillant at the origin, and find the directions of the tangents at that point.

(3) Show that $(y-x)^2 = x^3$ and $(y-x^2)^2 = x^5$ have cusps at the origin.

(4) Show that $(xy+1)^2 + (x-1)^3(x-2) = 0$ has a cusp at the point $x = 1$.

(5) Show that $x^4 - ax^2 y - axy^2 + a^2 y^2 = 0$ has a conjugate point at the origin.

(6) Find the singular points in the following curves:—

$$(y+x+1)^2 = (1-x)^5;$$

$$y^4 - axy^2 + x^4 = 0;$$

$$y^2 = x^3 - x^4;$$

$$y^4 + xy^3 + x^2(ax - by) = 0.$$

Contact of Curves.

237. Let
$$y = fx \text{ and } y = Fx$$
be the equations of two curves. If
$$fa = Fa,$$
the curves intersect at the point whose abscissa is a. If, in addition,
$$F'a = f'a,$$
the tangents at this point of intersection coincide, and the curves are said to have contact at the point in question. If
$$fa = Fa, \quad F'a = f'a, \text{ and } F''a = f''a,$$

the curves have contact of the second order at the point. If

$$Fa = fa, \quad F'a = f'a, \quad F''a = f''a, \text{ etc.}, \quad F^{(n)}a = f^{(n)}a,$$

the curves are said to have contact of the nth order at the point whose abscissa is a.

Contact of a higher order than the first is called *osculation*.

238. The difference between the ordinates of points of the two curves having the same abscissa and infinitely near the point of contact, is an infinitesimal of an order one higher than the order of contact of the curves.

Let
$$x = a + \varDelta x,$$
$$y_1 = f(a + \varDelta x),$$
and
$$y_2 = F(a + \varDelta x),$$

$$y_1 = fa + \varDelta x f'a + \frac{(\varDelta x)^2}{2!} f''a + \cdots + \frac{(\varDelta x)^n}{n!} f^{(n)}a$$

$$+ \frac{(\varDelta x)^{n+1}}{(n+1)!} f^{(n+1)}(a + \theta \varDelta x),$$

$$y_2 = Fa + \varDelta x F'a + \frac{(\varDelta x)^2}{2!} F''a + \cdots + \frac{(\varDelta x)^n}{n!} F^{(n)}a$$

$$+ \frac{(\varDelta x)^{n+1}}{(n+1)!} F^{(n+1)}(a + \theta' \varDelta x).$$

If the curves have contact of the nth order,

$$Fa = fa,$$

$$F'a = f'a, \text{ etc.}, \quad F^{(n)}a = f^{(n)}a.$$

$$y_1 - y_2 = \frac{(\varDelta x)^{n+1}}{(n+1)!} [f^{(n+1)}(a + \theta \varDelta x) - F^{(n+1)}(a + \theta' \varDelta x)],$$

CHAP. XVII.] THEORY OF PLANE CURVES. 253

which is infinitesimal of the $(n+1)$st order, if Δx is an infinitesimal. It follows, then, that the order of contact indicates the closeness of the contact; that is, the higher the order of contact of two curves, the less rapidly they recede from each other as they depart from the point of contact.

239. Let it be required to find the equation of the circle having contact of the second order with the curve

$$y = fx \qquad \text{at the point } (x_1, y_1).$$

Let a and b be the coördinates of the centre, and r the radius of the required circle. Call (X, Y) any point of the required circle, then its equation is

$$(X-a)^2 + (Y-b)^2 = r^2.$$

By our conditions,
$$\left(\frac{dY}{dX}\right)_{X=x_1} = \left(\frac{dy}{dx}\right)_{x=x_1},$$

$$\left(\frac{d^2Y}{dX^2}\right)_{X=x_1} = \left(\frac{d^2y}{dx^2}\right)_{x=x_1};$$

but
$$\frac{dY}{dX} = -\frac{X-a}{Y-b},$$

$$\left(\frac{dY}{dX}\right)_{X=x_1} = -\frac{x_1-a}{y_1-b}.$$

$$\frac{d^2Y}{dX^2} = \frac{-r^2}{(Y-b)^3},$$

$$\left(\frac{d^2Y}{dX^2}\right)_{X=x_1} = \frac{-r^2}{(y_1-b)^3},$$

hence
$$\left.\begin{aligned}\left(\frac{dy}{dx}\right)_{x=x_1} &= -\frac{x_1-a}{y_1-b} \\ \left(\frac{d^2y}{dx^2}\right)_{x=x_1} &= \frac{-r^2}{(y_1-b)^3}\end{aligned}\right\}$$

From these equations, and

$$(x_1 - a)^2 + (y_1 - b)^2 = r^2,$$

we can get the required values of a, b, and r.

Dropping accents, for the sake of simplicity,

$$y - b = -\left(\frac{r^2}{\frac{d^2y}{dx^2}}\right)^{\frac{1}{3}},$$

$$x - a = \left(\frac{r^2}{\frac{d^2y}{dx^2}}\right)^{\frac{1}{3}} \frac{dy}{dx};$$

substituting in $\quad (x - a)^2 + (y - b)^2 = r^2,$

$$\left(\frac{r^2}{\frac{d^2y}{dx^2}}\right)^{\frac{2}{3}} + \left(\frac{r^2}{\frac{d^2y}{dx^2}}\right)^{\frac{2}{3}}\left(\frac{dy}{dx}\right)^2 = r^2,$$

$$r^{\frac{2}{3}}\left(\frac{d^2y}{dx^2}\right)^{\frac{2}{3}} = 1 + \left(\frac{dy}{dx}\right)^2,$$

$$r\frac{d^2y}{dx^2} = \pm\left[1 + \left(\frac{dy}{dx}\right)^2\right]^{\frac{3}{2}},$$

$$r = \pm \frac{\left[1 + \left(\frac{dy}{dx}\right)^2\right]^{\frac{3}{2}}}{\frac{d^2y}{dx^2}},$$

which is the familiar value of the radius of curvature of

$$y = fx$$

at the point (x,y). Hence, our osculating circle is that circle having contact of the second order with the given curve at the point in question.

Examples.

(1) In the curve $y = x^4 - 4x^3 - 18x^2$,

show that the radius of curvature at the origin is $\frac{1}{36}$.

(2) Find the parabola whose axis is parallel to the axis of Y, which has the closest possible contact with the curve $y = \frac{x^3}{a^2}$ at the point where $x = a$. *Result.* $\left(x - \frac{a}{2}\right)^2 = \frac{a}{3}\left(y - \frac{a}{4}\right)$.

(3) Prove that, if the order of contact of two curves is even, they cross each other at the point of contact; if odd, they do not cross.

Envelops.

240. If the equation of a curve contain an undetermined constant, to different values of this constant will correspond different curves of a series. Such an equation is said to contain a *variable parameter*, the name being applied to a quantity which is constant for any one curve of a series, but varies in changing from one curve to another. For example: in the equation

$$(x - a)^2 + y^2 = r^2,$$

let a be a variable parameter; then the equation represents a series of circles, all having the radius r, and all having their centres on the axis of X.

A curve tangent to each of such a series of curves is called an *envelop*.

241. Two curves of such a series corresponding to two different values of the parameter will in general intersect. If they are made to approach each other indefinitely, by bringing the two values of the parameter nearer together, their point of intersection will evidently approach the enveloping curve, which then may be regarded as the *locus of the limiting position of a point*

of intersection of any two curves of the series as the curves are made to indefinitely approach. From this point of view the equation of an envelop is easily obtained.

Let
$$f(x,y,a) = 0 \qquad (1)$$

be the given equation of the series of curves, a being a variable parameter.
$$f(x,y,a+\Delta a) = 0 \qquad (2)$$

will be any second curve of the series. The equation

$$f(x,y,a+\Delta a) - f(x,y,a) = 0 \qquad (3)$$

represents some curve passing through all the points of intersection of (1) and (2) by the principle in Analytic Geometry: "If $u = 0$ and $v = 0$ are the equations of two curves, $u + kv = 0$ represents a curve containing all their points of intersection, and having no other point in common with them."

$$\frac{f(x,y,a+\Delta a) - f(x,y,a)}{\Delta a} = 0$$

is equivalent to (3). If, now, Δa be decreased indefinitely,

$$\lim_{\Delta a \doteq 0} \left[\frac{f(x,y,a+\Delta a) - f(x,y,a)}{\Delta a} \right] = 0,$$

or
$$D_a f(x,y,a) = 0, \qquad (4)$$

contains the limiting position of the point of intersection of (1) and (2). Let (x',y') be this point, and therefore any point of the required locus. Since (x',y') is on (4), and also on (1),

$$D_a f(x',y',a) = 0 \quad \text{and} \quad f(x',y',a) = 0;$$

we can eliminate a between these equations, and we shall have a single equation between x' and y', which will be the equation of the required envelop.

CHAP. XVII.]   THEORY OF PLANE CURVES.   257

242. For example: let us find the envelop of

$$(x - a)^2 + y^2 - r^2 = 0, \tag{1}$$

a being a variable parameter.

$$D_a f = -2(x - a) = 0.$$

$$x - a = 0. \tag{2}$$

Eliminating a between (1) and (2), we get

$$y^2 - r^2 = 0,$$

the equation of a pair of straight lines parallel to the axis of X, as the required envelop.

243. When dealing with the properties of evolutes, we proved that every normal to the original curve must be tangent to the evolute. We ought, then, to be able to find the evolute of any curve by treating it as the envelop of the normals of the curve.

Let $$y = fx$$

be the equation of the original curve

$$y - y_0 = -\left(\frac{dx}{dy}\right)_{x=x_0}(x - x_0)$$

is the equation of the normal, or

$$\left(\frac{dy_0}{dx_0}\right)(y - y_0) + x - x_0 = 0. \tag{1}$$

x_0 is the variable parameter,

$$D_{x_0} f = \frac{d^2 y_0}{dx_0^2}(y - y_0) - \left(\frac{dy_0}{dx_0}\right)^2 - 1 = 0, \tag{2}$$

$$y = y_0 + \frac{1 + \left(\frac{dy_0}{dx_0}\right)^2}{\frac{d^2 y_0}{dx_0^2}},$$

$$x = x_0 - \frac{\frac{dy_0}{dx_0}\left[1 + \left(\frac{dy_0}{dx_0}\right)^2\right]}{\frac{d^2y_0}{dx_0^2}};$$

but
$$y_0 = fx_0,$$

and we must eliminate x_0 and y_0, by the aid of these three equations, to obtain the equation of the evolute. These equations are the ones obtained by a different method in Art. 93.

EXAMPLES.

(1) Find the envelop of all ellipses having constant area, the axes being coincident.

Result. A pair of equilateral hyperbolas.

(2) A straight line of given length moves with its extremities on the two axes, required its envelop. *Result.* $x^{\frac{2}{3}} + y^{\frac{2}{3}} = a^{\frac{2}{3}}$.

(3) Find the envelop of straight lines drawn perpendicular to the normals to a parabola $y^2 = 4ax$ at the points where they cut the axis. *Result.* $y^2 = 4a(2a - x)$.

(4) Circles are described on the double ordinates of a parabola as diameters. Show that their envelop is an equal parabola.

(5) Find the envelop of all ellipses having the same centre, and in which the straight line joining the ends of the axes is of constant length. *Result.* $x \pm y = \pm c$.

(6) Show that the envelop of a circle on the focal radius of an ellipse as diameter is the circle on the major axis.

ADVERTISEMENTS

MATHEMATICAL TEXT-BOOKS.

For Higher Grades.

Anderegg and Roe:	Trigonometry	$0.75
Andrews:	Composite Geometrical Figures	.50
Baker:	Elements of Solid Geometry	.80
Beman and Smith:	Plane and Solid Geometry	1.25
Byerly:	Differential Calculus, $2.00; Integral Calculus	2.00
	Fourier's Series	3.00
	Problems in Differential Calculus	.75
Carhart:	Field-Book, $2.50; Plane Surveying	1.80
Comstock:	Method of Least Squares	1.00
Faunce:	Descriptive Geometry	1.25
Hall:	Mensuration	.50
Halsted:	Metrical Geometry	1.00
Hanus:	Determinants	1.80
Hardy:	Quaternions, $2.00; Analytic Geometry	1.50
	Differential and Integral Calculus	1.50
Hill:	Geometry for Beginners, $1.00; Lessons in Geometry	.70
Hyde:	Directional Calculus	2.00
Macfarlane:	Elementary Mathematical Tables	.75
Osborne:	Differential Equations	.50
Peirce (B. O.):	Newtonian Potential Function	1.50
Peirce (J. M.):	Elements of Logarithms, .50; Mathematical Tables	.40
Runkle:	Plane Analytic Geometry	2.00
Smith:	Coördinate Geometry	2.00
Taylor:	Elements of the Calculus	1.80
Tibbets:	College Requirements in Algebra	.50
Wentworth:	High School Arithmetic	1.00
	School Algebra, $1.12; Higher Algebra	1.40
	College Algebra	1.50
	Elements of Algebra, $1.12; Complete Algebra	1.40
	New Plane Geometry	.75
	New Plane and Solid Geometry	1.25
	Plane and Solid Geometry and Plane Trigonometry	1.40
	Analytic Geometry	1.25
	Geometrical Exercises	.10
	Syllabus of Geometry	.25
	New Plane Trigonometry	.40
	New Plane Trigonometry and Tables	.90
	New Plane and Spherical Trigonometry	.85
	New Plane and Spherical Trig. with Tables	1.20
	New Plane Trig. and Surveying with Tables	1.20
	New Plane and Spher. Trig., Surv., with Tables	1.35
	New Plane and Spher. Trig., Surv., and Navigation	1.20
Wentworth and Hill:	Exercises in Algebra, .70; Answers	.25
	Exercises in Geometry, .70; Examination Manual	.50
	Five-place Log. and Trig. Tables (7 Tables)	.50
	Five-place Log. and Trig. Tables (Complete Edition)	1.00
Wentworth, McLellan, and Glashan:	Algebraic Analysis	1.50
Wheeler:	Plane and Spherical Trigonometry and Tables	1.00

Descriptive Circulars sent, postpaid, on application.
The above list is not complete.

GINN & COMPANY, Publishers,
Boston. New York. Chicago. Atlanta. Dallas.

Wentworth's Mathematics

BY

GEORGE A. WENTWORTH, A.M.

ARITHMETICS

They produce practical arithmeticians.

Elementary Arithmetic	$ .30
Practical Arithmetic	.65
Mental Arithmetic	.30
Primary Arithmetic	.30
Grammar School Arithmetic	.65
High School Arithmetic	1.00
Wentworth & Hill's Exercises in Arithmetic (in one vol.)	.80
Wentworth & Reed's First Steps in Number.	
Teacher's Edition, complete	.90
Pupil's Edition	.30

ALGEBRAS

Each step makes the next easy.

First Steps in Algebra	$ .60
School Algebra	1.12
Higher Algebra	1.40
College Algebra	1.50
Elements of Algebra	1.12
Shorter Course	1.00
Complete	1.40
Wentworth & Hill's Exercises in Algebra (in one vol.)	.70

THE SERIES OF THE GOLDEN MEAN

GEOMETRIES

The eye helps the mind to grasp each link of the demonstration.

New Plane Geometry	$ .75
New Plane and Solid Geometry	1.25
P. & S. Geometry and Plane Trig.	1.40
Analytic Geometry	1.25
Geometrical Exercises	.10
Syllabus of Geometry	.25
Wentworth & Hill's Examination Manual	.50
Wentworth & Hill's Exercises	.70

TRIGONOMETRIES, ETC.

Directness of method secures economy of mental energy.

New Plane Trigonometry	$ .40
New Plane Trigonometry and Tables	.90
New Plane and Spherical Trig.	.85
New Plane and Spherical Trig. with Tables	1.20
New Plane Trig. and Surv. with Tables	1.20
Tables	.50 or 1.00
New P. and S. Trig., Surv., with Tables	1.35
New P. and S. Trig., Surv., and Nav.	1.20

The old editions are still issued.

GINN & COMPANY, Publishers,

Boston. New York. Chicago. Atlanta. Dallas.

Plane and Solid Geometry.

BY

WOOSTER WOODRUFF BEMAN,
Professor of Mathematics in the University of Michigan,

AND

DAVID EUGENE SMITH,
Professor of Mathematics in the Michigan State Normal College.

12mo. Half leather. ix + 320 pages. For introduction, $1.25.

While not differing radically from the standard American high school geometry in amount and order of material, this work aims to introduce and employ such of the elementary notions of modern geometry as will be helpful to the beginner.

Among these are the principles of symmetry, reciprocity or duality, continuity, and similarity.

The authors have striven to find a happy mean between books that leave nothing for the student to do and those that throw him entirely upon his own resources. At first the proofs are given with all detail; soon the references are given only by number; and finally much of the demonstration must be wrought out by the student himself.

The exercises have been carefully graded and have already been put to a severe test in actual practice. Effort has been made to guide in the solution of these exercises by a systematic presentation of the best methods of attacking "original" theorems and problems.

The leading text-books on geometry in English and other languages have been examined and their best features, so far as thought feasible, incorporated.

These merits will be found enhanced, it is believed, by the beauty and accuracy of the figures, and the excellence of the typographical make-up.

The publishers believe that this new Geometry will be cordially welcomed and that many teachers will find it admirably suited to the needs of their classes.

GINN & COMPANY, PUBLISHERS,

Boston. New York. Chicago. Atlanta. Dallas.

A TREATISE ON PLANE SURVEYING.

By **DANIEL CARHART**, C.E., *Professor of Civil Engineering in the Western University of Pennsylvania, Alleghany.*

Octavo. Half leather. 536 pages.   Introduction Price, $1.80.

This work covers the ground of Plane Surveying. It illustrates and describes the instruments employed, their adjustments and uses; it exemplifies the best methods of solving the ordinary problems occurring in practice, and furnishes solutions for many special cases which not infrequently present themselves. It is the result of twenty years' experience in the field and technical schools, and the aim has been to make it extremely practical, having in mind always that to become a *reliable* surveyor the student needs frequently to manipulate the various surveying instruments in the field, to solve many examples in the class-room, and to exercise good judgment in all these operations. Not only, therefore, are the different methods of surveying treated, and directions for using the instruments given, but these are supplemented by various field exercises to be performed, by numerous examples to be wrought, and by many queries to be answered.

Chapter I. Chain Surveying.
" II. Compass and Transit Surveying.
" III. Declination of the Needle.
" IV. Laying Out and Dividing Land.
" V. Plane Table Surveying.
" VI. Government Surveying.
" VII. City Surveying. Including the Principles of Levelling.
" VIII. Mine Surveying. Including the Elements of Topography.

A Table of Logarithms of Numbers; a Table of Natural and Logarithmic Sines, Cosines, Tangents, and Cotangents; a Traverse Table; and many others.

GINN & COMPANY, Publishers, Boston, New York, and Chicago.

PLANE AND SOLID

Analytic Geometry

By FREDERICK H. BAILEY, A.M. (Harvard), and FREDERICK S. WOODS, Ph.D. (Göttingen), Assistant Professors of Mathematics in Massachusetts Institute of Technology.

8vo. Cloth. 371 pages. For introduction, $2.00.

THIS book is intended for students beginning the study of analytic geometry, primarily for students in colleges and technical schools. While the subject-matter has been confined to that properly belonging to a first course, the treatment of all subjects discussed has been complete and rigorous. More space than is usual in text-books has been devoted to the more general forms of the equations of the first and the second degrees. The equations of the conic sections have been derived from a single definition, and after the simplest types of these equations have been deduced, the student is taught by the method of translation of the origin to handle any equation of the second degree in which the $x\,y$ term does not appear. In particular, the equations of the tangent, the normal, and the polar have been determined for such an equation. Only later is the general equation of the second degree fully discussed.

In the solid geometry, besides the plane and the straight line, the cylinders and the surfaces of revolution have been noticed, and all the quadric surfaces have been studied from their simplest equations. This study includes the treatment of tangent, polar, and diametral planes, conjugate diameters, circular sections, and rectilinear generators.

Throughout the work no use is made of determinants or calculus.

GINN & COMPANY, PUBLISHERS,
BOSTON.   NEW YORK.   CHICAGO.

BOTANIES

BOOKS OF SPECIAL VALUE.

ELEMENTS OF BOTANY.
By JOSEPH Y. BERGEN, Instructor in Biology in the English High School, Boston. 332 pages. Fully illustrated. For introduction, $1.10.

Bergen's Botany aims to revolutionize the study of botany and to put it on an experimental and observational basis, so that the study shall have a disciplinary value which it lacks now. The book can be used where they have no laboratory work, no microscope, in fact, no apparatus whatever. Good work can be done with a magnifying glass and pocket knife only. It covers a little more than a half year's work. The plan of the book is brought into substantial accord with the consensus of opinions of representative high school teachers in many sections of the country.

ELEMENTS OF PLANT ANATOMY.
By EMILY L. GREGORY, Professor of Botany in Barnard College. 148 pages. Illustrated. For introduction, $1.25.

Designed as a text-book for students who have already some knowledge of general botany, and who need a practical knowledge of plant structure.

ELEMENTS OF STRUCTURAL AND SYSTEMATIC BOTANY.
For High Schools and Elementary College Courses. By DOUGLAS H. CAMPBELL, Professor of Botany in the Leland Stanford Junior University. 253 pages. For introduction, $1.12.

PLANT ORGANIZATION.
By R. HALSTED WARD, formerly Professor of Botany in the Rensselaer Polytechnic Institute, Troy, N.Y. Quarto. 176 pages. Illustrated. Flexible boards. For introduction, 75 cents.

LITTLE FLOWER-PEOPLE.
By GERTRUDE E. HALE. Illustrated. 85 pages. For introduction, 40 cents.

This book tells some of the most important elementary facts of plant life in such a way as to appeal to the child's imagination and curiosity.

GLIMPSES AT THE PLANT WORLD.
By FANNY D. BERGEN. Fully illustrated. 156 pages. For introduction, 50 cents.

This is a capital child's book, and is intended for a supplementary reader for lower grades.

OUTLINES OF LESSONS IN BOTANY.
For the use of teachers or mothers studying with their children. By JANE H. NEWELL.
Part I.: **From Seed to Leaf.** 150 pages. Illustrated. For introduction, 50 cents.
Part II.: **Flower and Fruit.** 393 pages. Illustrated. For introduction, 80 cents.

A READER IN BOTANY.
Selected and adapted from well-known authors. By JANE H. NEWELL.
Part I.: **From Seed to Leaf.** 199 pages. For introduction, 60 cents.
Part II.: **Flower and Fruit.** 179 pages. For introduction, 60 cents.

Newell's Botanies aim to supply a course of reading in botany calculated to awaken the interest of pupils in the study of the life and habits of plants.

GINN & COMPANY, Publishers.

TEXT-BOOKS ON PHYSICS

By ALFRED P. GAGE,

Instructor in Physics in the English High School, Boston.

Principles of Physics. A text-book for high schools and colleges. 12mo. Half leather. 634 pages. Fully illustrated. For introduction, $1.30.

Introduction to Physical Science. Revised Edition. 12mo. Cloth. 374 pages. Illustrated. With a color chart of spectra, etc. For introduction, $1.00.

Elements of Physics. A text-book for high schools and academies. 12mo. Half leather. 424 pages. Illustrated. For introduction, $1.12.

Physical Laboratory Manual and Note Book. 12mo. Boards. 244 pages. Illustrated. For introduction, 35 cents.

Physical Experiments. A manual and note book. 12mo. Boards. 97 pages. Illustrated. For introduction, 35 cents.

THE **Principles of Physics** aims to supply the demand for an accurate, interesting, usable text-book of present-day physics, suitable for high schools and elementary courses in college.

The size and general features of the **Introduction to Physical Science** in its present revised form have been changed little, but numerous slight changes have been made throughout the work which will be found improvements and which will make it more acceptable to those using it.

The leading feature of the **Elements of Physics** is that it is strictly experiment-teaching in its method. The experiments given are rather of the nature of questions than of illustrations, and precede the statements of principles and laws.

Physical Experiments contains the laboratory exercises required for admission to Harvard University and to many other colleges. Specific directions are given for the preparation of notes, thereby securing uniformity which greatly reduces the labor of the examiner.

GINN & COMPANY, Publishers,

Boston. New York. Chicago. Atlanta. Dallas.

TEXT-BOOKS ON CHEMISTRY

BY R. P. WILLIAMS,
Instructor in Chemistry in the English High School, Boston.

Elements of Chemistry. 12mo. Cloth. 412 pages. Fully illustrated. For introduction, $1.10.

Introduction to Chemical Science. 12mo. Cloth. 216 pages. Illustrated. For introduction, 80 cents.

Chemical Experiments. General and Analytical. 8vo. Boards. 212 pages. Fully illustrated. For introduction, 50 cents.

Laboratory Manual of Inorganic Chemistry. One hundred topics in general, qualitative, and quantitative chemistry. 12mo. Boards. 200 pages. Illustrated. For introduction, 30 cents.

Williams' Laboratory Manual of General Chemistry is still kept in stock. For introduction, 25 cents.

THE **Elements of Chemistry** is very fully and carefully illustrated with entirely new designs, embodying many original ideas, and there is a wealth of practical experiments. Exercises and problems follow the discussion of laws and principles.

The subject-matter is so divided that the book can be used by advanced schools, or by elementary ones in which the time allotted to chemistry is short.

Chemical Experiments is for the use of students in the chemical laboratory. It contains more than one hundred sets of the choicest illustrative experiments, about half of which belong to general chemistry, the rest to metal and acid analysis.

The **Laboratory Manual** contains one hundred sets of experiments in inorganic general chemistry, including brief treatment of qualitative analysis of both metals and non-metals, and a few quantitative experiments.

GINN & COMPANY, Publishers,

Boston. New York. Chicago. Atlanta. Dallas.

ASTRONOMY

By CHARLES A. YOUNG, Ph.D., LL.D., Professor of Astronomy in Princeton University, Princeton, N.J., and author of *The Sun*.

A Series of text-books on astronomy for high schools, academies, and colleges. Prepared by one of the most distinguished astronomers of the world, a most popular lecturer, and a most successful teacher.

Lessons in Astronomy. Including Uranography. Revised Edition. 12mo. Cloth. Illustrated. 366 pages, exclusive of four double-page star maps. For introduction, $1.20.

Elements of Astronomy. With a Uranography. 12mo. Half leather. 472 pages, and four star maps. For introduction, $1.40.

    Uranography. From the *Elements of Astronomy*. Flexible covers. 42 pages, besides four star maps. For introduction, 30 cents.

General Astronomy. A text-book for colleges and technical schools. 8vo. 551 pages. Half morocco. Illustrated with over 250 cuts and diagrams, and supplemented with the necessary tables. For introduction, $2.25.

The **Lessons in Astronomy** (recently brought up to date) was prepared for schools that desire a brief course free from mathematics. Everything has been carefully worked over and rewritten to adapt it to the special requirements. Great pains has been taken not to sacrifice accuracy and truth to brevity, and no less to bring everything thoroughly up to the present time.

The **Elements of Astronomy** is an independent work, and not an abridgment of the author's General Astronomy. It is a text-book for advanced high schools, seminaries, and brief courses in colleges generally. Special attention has been paid to making all statements correct and accurate so far as they go.

The eminence of Professor Young as an original investigator in astronomy, a lecturer and writer on the subject, and an instructor in college classes, led the publishers to present the **General Astronomy** with the highest confidence; and this confidence has been fully justified by the event. It is conceded to be the best astronomical text-book of its grade to be found anywhere.

GINN & COMPANY, Publishers,
Boston. New York. Chicago. Atlanta. Dallas.

THE BEST HISTORIES.

Myers's History of Greece. — Introduction price, $1.25.

Myers's Eastern Nations and Greece. — Introduction price, $1.00.

Allen's Short History of the Roman People. — Introduction price, $1.00.

Myers and Allen's Ancient History. — Introduction price, $1.50. This book consists of Myers's Eastern Nations and Greece and Allen's History of Rome bound together.

Myers's History of Rome. — Introduction price, $1.00.

Myers's Ancient History. — Introduction price, $1.50. This book consists of Myers's Eastern Nations and Greece and Myers's History of Rome bound together.

Myers's Mediæval and Modern History. — Introduction price, $1.50.

Myers's General History. — Introduction price, $1.50.

Emerton's Introduction to the Study of the Middle Ages. — Introduction price, $1.12.

Emerton's Mediæval Europe (814-1300). — Introduction price, $1.50.

Fielden's Short Constitutional History of England. — Introduction price, $1.25.

Montgomery's Leading Facts of English History. — Introduction price, $1.12.

Montgomery's Leading Facts of French History. — Introduction price, $1.12.

Montgomery's Beginner's American History. — Introduction price, 60 cents.

Montgomery's Leading Facts of American History. — Introduction price, $1.00.

Cooper, Estill and Lemmon's History of Our Country. — Introduction price, $1.00.

For the most part, these books are furnished with colored and sketch maps, illustrations, tables, summaries, analyses and other helps for teachers and students.

GINN & COMPANY, Publishers.

BOSTON.   NEW YORK.   CHICAGO.   ATLANTA.

INDUCTIVE LOGIC

By WM. G. BALLANTINE,
President of Oberlin College.

12mo. Cloth. viii + 174 pages. For introduction, 80 cents.

ALTHOUGH this is preëminently a scientific age, surprisingly little attention is given in our colleges and universities to the study of inductive logic. The neglect is probably due to the lack of a satisfactory text-book. Such manuals as are found are hardly more than meager abridgments of some chapters of the great but unequal work of Mill.

It is the aim of this book to present, within reasonable compass, a fresh and independent statement of the fundamental principles of inductive logic, consistently carried out in detail and amply illustrated by extracts from a wide range of philosophical and scientific writers. The best modern teachers make large use of the library and, while setting forth their own views, seek to acquaint their students with the literature of the subject and the history of opinion. It is believed that these numerous quotations from Bacon, Mill, Darwin, Helmholtz, G. F. Wright, and others, while exactly in point as illustrations and elucidations, will also be found strikingly interesting in themselves and highly useful in familiarizing the reader with the phraseology, literary styles and modes of thinking of those eminent authorities.

Teachers of inductive logic will be pleased to find here a simple account of the relations of induction and deduction which discards the notion of two separate realms of thought in one of which it is, and in the other is not, legitimate to draw a conclusion wider than the premises. The classification of inductions under three heads, as primary, secondary and mixed, clears away the confusions which have arisen from the attempt to bring all inductions under a single definition. The doctrine of Causation is treated with great thoroughness, but the notion of cause is not made, as in Mill's system, the root of the whole theory of induction.

GINN & COMPANY, PUBLISHERS,
Boston.   New York.   Chicago.   Atlanta.

Glaciers of North America

By ISRAEL C. RUSSELL,

*Professor of Geology in the University of Michigan,
Author of "Lakes of North America."*

8vo. Cloth. x + 210 pages. Illustrated.
By mail, postpaid, $1.90; to teachers and for introduction, $1.75.

RECENT explorations have shown that North America contains thousands of glaciers, some of which are not only vastly larger than any in Europe, but belong to types of ice bodies not there represented. In the study of the glaciers of North America, and especially of those in Alaska, Professor Russell has taken an active part, and this book not only presents the results of his own explorations, but a condensed and accurate statement of the present status of glacial investigations. Its popular character and numerous illustrations will make it of interest to the general reader.

Edwin S. Balch, *Vice-President Geographical Society of Philadelphia:* It is by all odds the clearest work on glaciers in general that I have ever seen.

F. Bascom, *Department of Geology, Bryn Mawr College, Pa.:* No one is better fitted than Professor Russell, by extended personal explorations, to discuss with authority this subject. His volume combines in an unusual degree a clearness and fascination of style with a thoroughly scientific treatment.

Henry F. Osborne, *American Museum of Natural History, New York, N.Y.:* I consider it a contribution of very great value. By bringing together all that is at present known upon the glaciers of North America in convenient form, this work will stimulate the study of our own living glaciers among American geological students and thus accomplish a marked service to American geology.

Department of Special Publication.

GINN & COMPANY, PUBLISHERS,

Boston, New York, and Chicago.

www.ingramcontent.com/pod-product-compliance
Lightning Source LLC
Chambersburg PA
CBHW032100220426
43664CB00008B/1077